The Ultimate Guide to Snowpark

Design and deploy Snowflake Snowpark with Python
for efficient data workloads

Shankar Narayanan SGS

Vivekanandan SS

The Ultimate Guide to Snowpark

Group Product Manager: Kaustubh Manglurkar
Publishing Product Manager: Apeksha Shetty
Book Project Manager: Hemangi Lotlikar
Content Development Editor: Manikandan Kurup
Technical Editor: Seemanjay Ameriya
Copy Editor: Safis Editing
Proofreader: Manikandan Kurup
Indexer: Hemangini Bari
Production Designer: Joshua Misquitta
Senior DevRel Marketing Executive: Nivedita Singh

First published: May 2024
Production reference: 2190924

Published by Packt Publishing Ltd.
Grosvenor House
11 St Paul's Square
Birmingham
B3 1RB, UK

ISBN 978-1-80512-341-5
www.packtpub.com

To my caring mother, Geetha, and inspiring father, Ganapathy, who taught me to aim high. To my lovely sister, Deepa, who tolerated all my pranks yet continued to love me unconditionally.

To my life partner, Amrita – thank you for being my loving partner and motivating and supporting me in completing the book.

To Snowflake for creating the comprehensive data cloud platform and to all the technologists and developers whose work inspired the book.

– Shankar Narayanan SGS

I dedicate this book to the loving memory of my cherished parents, the Late S.R. Srinivasan and the Late S. Chandra. Their legacy of love and strength continues to shape my journey each day.

I am deeply grateful for the unwavering support and love of my brother, S.S. Sathish Kumar, and sister-in-law, R. Ranjani. They have stood by my side steadfastly through every high and low of my life.

My heartfelt thanks go to my beloved aunt, R. Rani, and to my little bundle of joy, Skandhaguru (Skandhu), who has loved me no matter what through this journey.

Finally, I extend my heartfelt gratitude to the remarkable transplant team at Kauvery Hospital, Chennai, for guiding me through my renal transplant journey, even as I write and publish this book.

– Vivekanandan SS

Foreword

This is a unique and critical juncture point in the industry. Data plays an increasingly important role in every organization. The volumes and ecosystems of data continue to grow. AI and ML will continue to define and redefine business models and customer experiences. To complement these forces is an emerging set of tools and technologies to help you explore these exciting new worlds.

With that framing in mind, I'm thrilled to introduce *The Ultimate Guide to Snowpark: Design and deploy Snowflake Snowpark with Python for efficient data workloads*, authored by the Data Superhero Shankar Narayanan SGS and his co-author Vivekanandan SS. This book arrives at a critical juncture in the evolution of data processing, where efficiency and scalability are paramount, and Snowpark stands as a beacon of innovation within Snowflake's ecosystem.

The core premise of Snowpark is simple: how can we improve data processing tasks and data applications by moving the code and processing to the data? Snowpark represents a transformative approach to managing data workloads, offering developers and data scientists the tools to build robust, scalable data solutions directly within Snowflake. By integrating Python, Snowpark unleashes the power of one of the most popular programming languages, making it accessible not just to data engineers but also to a broader community of technologists seeking to leverage data in new and powerful ways.

This guide is more than just a technical manual; it is a gateway to mastering Snowpark. The authors have meticulously crafted a resource that balances foundational knowledge with advanced techniques, providing clear, actionable guidance complemented by practical examples and code snippets. Their deep understanding of Snowpark's capabilities is evident in every chapter, making this book an indispensable tool for anyone eager to excel in the Snowflake environment.

As the Director of Product at Snowflake managing Snowpark, I am profoundly proud and grateful to the authors who invested their expertise and passion into empowering our user community. This book does an exceptional job of demystifying complex concepts and lays down a roadmap for deploying sophisticated data engineering and science projects that are both innovative and practical.

To the readers, whether you are starting your journey with Snowflake or looking to expand your existing skills, *The Ultimate Guide to Snowpark* will equip you with the knowledge to transform your ideas into reality, elevate your projects, and lead the way in data-driven innovation.

Enjoy the read, and may it inspire you to push the boundaries of what is possible with Snowpark and Snowflake. I can't wait to see what you can create, build, and unlock.

Jeff Hollan

Director of Product, Snowflake

Contributors

About the authors

Shankar Narayanan SGS is a principal architect at Microsoft, with over a decade of diverse experience leading and delivering large-scale data and cloud implementations for Fortune 500 companies across various industries. He has successfully implemented the Snowflake Data Cloud for many organizations, leading customers to adopt Snowflake.

He holds bachelor's and master's degrees in computer science and many certifications in multi-cloud platforms and Snowflake. He is an award-winning blogger, contributing to various technical publications and open source projects. He has been selected as the SAP community topic leader. He has been chosen as one of the Snowflake Data Heroes by Snowflake and the recipient of a Top 10 Data and Analytics Professional award by OnCon.

Vivekanandan SS spearheads the GenAI enablement team at Verizon, leveraging over a decade of expertise in data science and big data. His professional journey spans across building analytics solutions and products across diverse domains, and proficient in cloud analytics and data warehouses.

He holds a bachelor's degree in industrial engineering from Anna University, a long-distance program in big data analytics from IIM, Bangalore, and a master's in data science from Eastern University. As a seasoned trainer, he imparts his knowledge, specializing in Snowflake and GenAI, and is also a data science guest faculty and advisor for various educational institutes. His solution is ranked in the top 1 percentile in Kaggle kernels globally.

About the reviewers

Preston Blackburn is a machine learning engineer with a background in data engineering. He has worked at multiple start-ups, specializing in Snowflake consulting and built libraries that extend Snowpark functionality. Preston excels in developing internal developer tools, accelerating data modernization efforts, and architecting robust ML pipelines. His dedication to pushing the boundaries of technology drives innovation and ensures his clients stay at the forefront of industry advancements.

Balamurugan Kannaiyan is a highly accomplished data engineering leader, with around two decades of experience specializing in cloud technologies (AWS, Azure, and Snowflake), data management, and advanced analytics.

He currently leads the data engineering team at a Texas Public Sector, leveraging Snowflake's cutting-edge capabilities to build high-performance data applications. Bala brings a distinguished track record from prior roles within the public sector and Fortune 500 companies in designing scalable, cloud-distributed systems. Further solidifying his expertise, Bala holds a bachelor of engineering degree from the prestigious Anna University alongside numerous certifications and accreditations in Snowflake, Azure, Databricks, and Oracle.

Table of Contents

Part 2: Snowpark Data Workloads

3

4

5

6

Deploying and Managing ML Models with Snowpark   149

Part 3: Snowpark Applications

7

Developing a Native Application with Snowpark   177

8

Introduction to Snowpark Container Services 203

Preface

Snowpark is a powerful framework that helps you unlock a realm of possibilities within the Snowflake Data Cloud. However, leveraging Snowpark's full potential with Python can be challenging without proper guidance. Packed with practical examples and code snippets, this book will guide you to successfully use Snowpark with Python.

The Ultimate Guide to Snowpark helps you develop an understanding of Snowpark and how it integrates various workloads, such as data engineering, data science, and data applications, within the Data Cloud.

From configuration to coding styles and workloads such as data manipulation, collection, preparation, transformation, aggregation, and analysis, this guide will equip you with the right knowledge to make the most of this framework. You'll discover how to build, test, and deploy data pipelines and data science models. As you progress, you'll deploy data applications natively in Snowflake and operate with LLMs using Snowpark container services.

By the end, you'll be well-positioned to leverage Snowpark's capabilities and propel your career as a Snowflake developer to new heights.

Who this book is for

This book is for data engineers, data scientists, data architects, application developers, and data practitioners seeking an in-depth understanding of Snowpark features and best practices for deploying various workloads in Snowpark using the Python programming language.

What this book covers

Chapter 1, Discovering Snowpark, will guide you through Snowpark and its unique capabilities. You will learn how to utilize Python with Snowpark and how to implement it for various workloads. By the end of this chapter, you will grasp Snowpark's functionalities and benefits, including faster data processing, improved data quality, and reduced costs. These guided chapters aim to give you an all-encompassing understanding of Snowpark and how to leverage its value for your specific use cases.

Chapter 2, Establishing a Foundation with Snowpark, teaches you how to configure and operate Snowpark, establish coding style and structure, and explore workloads. You will also acquire practical knowledge and skills to work efficiently with Snowpark, including setting up the environment, structuring the code, and utilizing it for different workloads.

Chapter 3, *Simplifying Data Processing Using Snowpark*, teaches users how to work with data in Snowpark. It covers data collection, preparation, transformation, aggregation, and analysis. By the end of this chapter, users will gain practical knowledge and skills for managing data sources, cleaning and transforming data, and performing advanced analysis tasks.

Chapter 4, *Building Data Engineering Pipelines with Snowpark*, covers building reliable data pipelines, effective debugging and logging, efficient deployment using DataOps, and test-driven development for Snowpark. This chapter will equip users with practical skills for developing, testing, and deploying data pipelines, resulting in reliable and efficient channels in Snowpark.

Chapter 5, *Developing Data Science Projects with Snowpark*, covers the use of Snowpark for data science projects, as well as exploring the data science pipeline, which includes data preparation, exploration, and model training featuring Snowpark. The material caters to data scientists and other professionals looking to use Snowpark to tackle extensive data processing and construct precise machine-learning models.

Chapter 6, *Deploying and Managing ML Models with Snowpark*, explores implementing machine learning models in Snowpark and constructing a feature store. Additionally, readers can learn to integrate model registry into Snowpark and monitor and operationalize their ML models. This chapter caters to data scientists and experts who aspire to master the techniques of deploying and administering their machine-learning models competently with Snowpark.

Chapter 7, *Developing a Native Application with Snowpark*, will explore the Native App Framework and how to develop, deploy, manage, and monetize a Native App using Snowpark. This chapter caters to developers who aspire to build apps within Snowflake.

Chapter 8, *Introduction to Snowpark Container Services*, introduces Snowpark Container Services and discusses how to deploy applications in containers within Snowflake. This chapter caters to developers building container applications in Snowflake.

To get the most out of this book

To benefit fully from this book, you should have basic knowledge of SQL, proficiency in Python, a grasp of data engineering and data science basics, and familiarity with the Snowflake Data Cloud platform.

Software/hardware covered in the book	Operating system requirements
Snowflake	Windows, macOS, or Linux (any)
Python	Windows, macOS, or Linux (any)
Visual Studio Code	Windows, macOS, or Linux (any)
Chrome or another latest Browser	

You will need a Snowflake account. You can sign up for a trial at https://signup.snowflake.com/.

If you are using the digital version of this book, we advise you to type the code yourself or access the code from the book's GitHub repository (a link is available in the next section). Doing so will help you avoid any potential errors related to the copying and pasting of code.

Download the example code files

You can download the example code files for this book from GitHub at `https://github.com/PacktPublishing/The-Ultimate-Guide-To-Snowpark`. If there's an update to the code, it will be updated in the GitHub repository.

We also have other code bundles from our rich catalog of books and videos available at `https://github.com/PacktPublishing/`. Check them out!

Conventions used

There are a number of text conventions used throughout this book.

`Code in text`: Indicates code words in text, database table names, folder names, filenames, file extensions, pathnames, dummy URLs, user input, and Twitter handles. Here is an example: "An event table called `MY_EVENTS` is created with the default column structure."

A block of code is set as follows:

```
current_runs = dag_op.get_current_dag_runs(dag)
for r in current_runs:
    print(f"RunId={r.run_id} State={r.state}")
```

Any command-line input or output is written as follows:

```
conda create --name def_gui_3.8_env --override-channels --channel
https://repo.anaconda.com/pkgs/snowflake python=3.8
```

Bold: Indicates a new term, an important word, or words that you see onscreen. For instance, words in menus or dialog boxes appear in **bold**. Here is an example: "A **model registry** is a centralized repository that enables model developers to organize, share, and publish ML models efficiently."

> Tips or important notes
> Appear like this.

Get in touch

Feedback from our readers is always welcome.

General feedback: If you have questions about any aspect of this book, email us at customercare@ packtpub.com and mention the book title in the subject of your message.

Errata: Although we have taken every care to ensure the accuracy of our content, mistakes do happen. If you have found a mistake in this book, we would be grateful if you would report this to us. Please visit www.packtpub.com/support/errata and fill in the form.

Piracy: If you come across any illegal copies of our works in any form on the internet, we would be grateful if you would provide us with the location address or website name. Please contact us at copyright@packt.com with a link to the material.

If you are interested in becoming an author: If there is a topic that you have expertise in and you are interested in either writing or contributing to a book, please visit authors.packtpub.com.

Share your thoughts

Once you've read *The Ultimate Guide to Snowpark*, we'd love to hear your thoughts! Scan the QR code below to go straight to the Amazon review page for this book and share your feedback.

https://packt.link/r/1-805-12341-6

Your review is important to us and the tech community and will help us make sure we're delivering excellent quality content.

Download a free PDF copy of this book

Thanks for purchasing this book!

Do you like to read on the go but are unable to carry your print books everywhere?

Is your eBook purchase not compatible with the device of your choice?

Don't worry, now with every Packt book you get a DRM-free PDF version of that book at no cost.

Read anywhere, any place, on any device. Search, copy, and paste code from your favorite technical books directly into your application.

The perks don't stop there, you can get exclusive access to discounts, newsletters, and great free content in your inbox daily

Follow these simple steps to get the benefits:

1. Scan the QR code or visit the link below

https://packt.link/free-ebook/9781805123415

2. Submit your proof of purchase
3. That's it! We'll send your free PDF and other benefits to your email directly

Part 1: Snowpark Foundation and Setup

In this part, we will explore the fundamental and advanced features of the Snowpark framework in Python. This part focuses on the Snowpark platform and the setup required to get started using Snowpark.

This part includes the following chapters:

- *Chapter 1, Discovering Snowpark*
- *Chapter 2, Establishing a Foundation with Snowpark*

1

Discovering Snowpark

Snowpark is the recent major innovation released by Snowflake that provides an intuitive set of libraries and runtimes for querying and processing data at scale in Snowflake. This chapter aims to guide you through Snowpark to understand its unique capabilities. In addition, the chapter helps you learn how to utilize Python with Snowpark and implement it in various workloads such as data engineering, data science, and data applications. By the end of this chapter, you will have grasped Snowpark's capabilities and benefits, including faster data processing, scalability, and reduced costs.

In this chapter, we're going to cover the following main topics:

- Introducing Snowpark
- Leveraging Python for Snowpark
- Understanding Snowpark for different workloads
- Realizing the value of using Snowpark

Introducing Snowpark

Snowflake, founded in 2012, started its journey to the data cloud by completely re-engineering the world of data and rethinking how a reliable, secure, high-performance, and scalable data-processing system should be architected for the cloud. It started with offering cloud-based data warehousing through a managed **Software as a Service (SaaS)** platform to load, analyze, and process large volumes of data. The success of Snowflake lies in the fact that it is a cloud-native managed solution that is built on top of the major public cloud providers such as Amazon Web Services, Microsoft Azure, and Google Cloud Platform by automatically providing a reliable, secure, high-performance, and scalable data processing system for organizations without the need to deploy hardware or install or configure any software.

As with any cloud data warehousing, Snowflake supports **American National Standards Institute (ANSI)** SQL as the language of choice. Although SQL is a powerful declarative language that allows users to ask questions about data, it is constrained to data warehouse workloads, limiting the support for advanced workloads such as data science and data engineering, which require developers to write the solution in other programming languages leading them to move data out of Snowflake to perform these workloads.

Snowflake's solution to this challenge is **Snowpark**, an innovative developer framework that streamlines the process of building complex data pipelines. With Snowpark, data scientists and developers can directly interact with Snowflake using their preferred programming language, enabling them to quickly and securely deploy **machine learning** (**ML**) models, execute data pipelines, and develop data applications on Snowflake's virtual compute warehouse in a serverless manner without having to transfer data outside of Snowflake.

Snowpark enables data teams to collaborate on the data by natively supporting work with DataFrame style programming in Python, Scala, or Java, exposing deeply integrated interfaces in these languages to augment Snowflake's original SQL language and minimizing the complexity of having to manage different environments for advanced data pipelines. This has led developers to leverage Snowflake's robust and scalable computing power to ship code to the data without exporting it outside Snowflake into other environments.

In this section, we covered a brief introduction to Snowpark and learned how it fits into the Snowflake ecosystem and how it helps developers. The following section will cover how to leverage Python for Snowpark.

Leveraging Python for Snowpark

In June 2022, Snowflake made a significant announcement, revealing the much-anticipated **Snowpark for Python**. This new release has rapidly emerged as the preferred programming language for Snowpark, providing users with a more extensive range of options for programming data in Snowflake. Moreover, Snowpark has simplified managing data architectures, enabling users to operate more quickly and efficiently.

Snowpark for Python is a cutting-edge, enterprise-grade, open-source innovation integrated into the Snowflake data cloud. As a result, the platform delivers a seamless, unified experience for data scientists and developers. In addition, the Snowpark for Python package is built upon the Snowflake Python connector. The Python connector enables users to execute SQL commands and other essential functions in Snowflake and Snowpark for Python empowers users to undertake more advanced data applications.

For instance, the platform permits users to run **user-defined functions** (**UDFs**), **external functions**, and **stored procedures** directly within Snowflake. This powerful new functionality enables data scientists, engineers, and developers to create robust and secure data pipelines and ML models within Snowflake. As a result, they can leverage the platform's superior performance, elasticity, and security features to deliver advanced insights and drive meaningful business outcomes. Overall, Snowpark for Python represents a significant step forward for Snowflake, offering users enhanced functionality and flexibility while retaining the platform's exceptional performance and security features.

Snowpark for Python supports pre-vetted open-source packages through integration with the **Anaconda** environment that executes on an Anaconda-powered sandbox inside Snowflake's virtual compute warehouses, which provides a familiar interface for the developers. The integrated Anaconda package manager is valuable for developers as it comes with a comprehensive set of curated open-source packages and supports resolving dependencies between different packages and versions. It is a huge time-saver and helps prevent developers from dealing with "dependency hell."

Capabilities of Snowpark for Python

Snowpark for Python is generally available across all cloud instances of Snowflake. It helps accelerate different workloads and comes with a rich set of capabilities, as follows:

- It allows developers to write Python code within Snowflake, enabling them to directly leverage the power of Python libraries and frameworks in Snowflake

- It supports popular open-source Python libraries such as pandas, NumPy, SciPy, and scikit-learn, along with other libraries, allowing developers to perform complex data analysis and ML tasks directly within Snowflake

- It also provides access to external data sources such as AWS S3, Azure Blob storage, and Google Cloud Storage, allowing developers to work with data stored outside Snowflake

- It provides seamless integration with Snowflake's SQL engine, allowing developers to write queries using functional programming methods with Python that compile to SQL

- It also supports distributed processing, allowing developers to scale their Python code to handle large datasets and complex logic

- It enables developers to build custom UDFs that can be used within SQL queries, allowing for greater flexibility and customization of data processing workflows

- Snowpark provides a Python development environment within Snowflake, allowing developers to write, test, and debug Python code directly within the Snowflake UI

- It enables developers to work with various data formats such as CSV, JSON, Parquet, and Avro, providing data processing and analysis flexibility

- It provides a unified data processing experience that works with SQL and Python in a single environment

- It enables developers to create custom data pipelines using Python code, making integrating Snowflake with other data sources and data processing tools easier

- It can handle real-time and batch data processing, making it easier to build data-intensive workloads

- It provides a robust framework built on Snowflake that ensures data privacy and compliance with industry standards such as the **Health Insurance Portability and Accountability Act (HIPAA)**, **General Data Protection Regulation (GDPR)**, and **Security Operations Center (SOC)**

- Snowpark supports enhancing data by leveraging **Snowflake Marketplace**

Snowpark for Python packs many capabilities that help developers use it efficiently for various workloads and use cases within Snowflake.

Why Python for Snowpark

Although Snowpark supports Python, Scala, and Java, this book will focus only on Python, a de facto for Snowpark development. Python's growing popularity through high-level built-in data structures with dynamic typing and binding makes it ideal for data operations. In addition, the language is very flexible and easy to learn by developers. Its power lies in the rich open-source ecosystem that is well-supported with a growing list of popular packages.

Python is a general-purpose, versatile programming language for different purposes, such as data engineering, data science, and data applications. It enables developers to learn a single programming language for all their needs.

Snowflake is also heavily investing in Python to make it easier for data scientists, engineers, and application developers to build even more in the data cloud without governance trade-offs.

In this section, we covered the capabilities of Snowpark for Python and why Python is a preferred language for developing Snowpark. The following section will cover how Snowpark can be used for different workloads.

Understanding Snowpark for different workloads

The release of Snowpark transformed Snowflake into a complete data platform designed to support various workloads. Snowpark supports multiple workloads, such as data science and ML, data engineering, and data applications.

Data science and ML

Python is the favorite language for data scientists. Snowpark for Python supports popular libraries and frameworks such as pandas, NumPy, and scikit-learn, making it the ideal framework for data scientists to perform ML development in Snowflake. In addition, data scientists can use the DataFrames API to interact with data inside Snowflake and perform batch training and inference inside Snowflake. Developers can also use Snowpark for feature engineering, ML model inference, and end-to-end ML pipelines. Snowpark also provides a SnowparkML library to support data science and ML in Snowpark.

Data engineering

Data cleansing and ELT workloads are complex, and building a data pipeline with just SQL is where Snowpark can be of great benefit. Snowpark lets developers factor code for readability and reuse it while providing a better capability for unit tests. In addition, with the support of Anaconda, developers can use open-source Python libraries for building reliable data pipelines. The other major challenge with data processing is that the infrastructure requires significant manual effort and maintenance. Snowpark solves this problem by being highly performant, enabling data engineers to work with large datasets quickly and efficiently, building complex data pipelines, and processing large volumes of data without performance issues.

Data governance and security

Snowpark supports developing solutions that incorporate data governance and security. Data governance is critical and augments the data science and data engineering use cases. Snowpark simplifies the governance posture by helping organizations understand and improve data quality. Developers can quickly create a function to perform data tests and detect anomalies. Snowpark can utilize the data classification capability to detect **personally identifiable information** (**PII**) and classify data that is critical to an organization. Custom functions developed in Snowpark can mask sensitive data such as credit card numbers using the robust dynamic data masking feature while retaining the existing security model in Snowflake.

Data applications

Snowpark helps the team develop dynamic data applications that run directly on Snowflake without moving the data outside. Using **Streamlit**, a powerful open-source library that Snowflake acquired, developers can build native applications using the familiar Python environment. Interactive ML-powered applications can be developed and shared with users securely utilizing role-based access controls entirely on Snowflake's governed platform, taking advantage of its scale, performance, and governance. The Snowflake Native Application Framework provides a streamlined path to monetize apps through Snowflake Marketplace, where you can make your app available to other Snowflake customers and open new revenue opportunities.

Snowpark supports different workloads and makes Snowflake a complete data cloud solution. The following section will highlight Snowpark's technical and business benefits.

Realizing the value of using Snowpark

The traditional big data approach has been in the industry for a long time and is unsuitable for modern cloud-based scalable workloads. Traditional architecture has many challenges, such as the following:

- De-coupling the compute and data into separate systems
- Running separate processing clusters for different languages
- Complexity in managing the system
- Data silos and data duplication
- Lack of unified security and governance

Snowflake solves the traditional system's challenges using Snowpark, providing tremendous value to the data ecosystem and Snowflake users. The following diagram shows the difference between a traditional approach and Snowflake's streamlined approach:

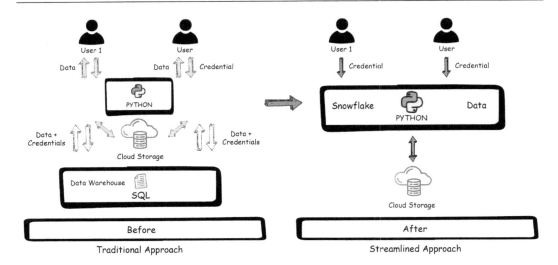

Figure 1.1 – Traditional versus Snowflake approach

As you can see from the difference between both approaches, Snowpark's streamlined approach benefits both the business and the developers by providing a flexible, efficient, and cost-effective way to build data that scales with the business needs. Some of the significant values of using Snowpark are as follows:

- Snowpark can access data programmatically through the DataFrame APIs, making the data ingestion and integration consistent, as you can integrate various structured and unstructured data

- Snowpark standardizes the approach to data processing since the data pipelines are in Python code; they can be tested and deployed and are easier to understand and interpret

- Anaconda powers Snowpark for Python and provides easy access to third-party Python libraries that are open source, which enhances the data processing capabilities and empowers developers to perform more

- Snowpark integrates and runs seamlessly on the existing Snowflake virtual warehouse, allowing developers to build data applications designed to scale without any additional infrastructure

- Snowpark's framework supports various workloads, such as data engineering, data science, and data applications, providing a unified experience for development on the data cloud

- Snowpark delivers a secure, governed environment as it is easy to enforce governance policies, and there is no data movement outside Snowflake

Let's wrap up this chapter.

Summary

Snowflake's Snowpark perfectly coalesces SQL and Python, running complex data processing jobs in the Snowflake data cloud and enabling data engineers, data scientists, and developers to take advantage of Snowflake. In this chapter, we have seen the benefits of Snowpark and why Python is the preferred development language. We also covered different workloads that Snowpark supports.

In the next chapter, we will examine configuring and operating with Snowpark in detail and learn how to use Snowpark for various workloads.

2

Establishing a
Foundation with Snowpark

In the previous chapter, you learned the basics of Snowpark, its benefits, and how it allows developers to develop complex data applications using Python. This chapter will focus on establishing a solid foundation with Snowpark. Here, you will learn how to configure and operate Snowpark, select a coding style and structure, and explore Snowpark's fundamentals in depth. This will help you acquire practical knowledge and skills to work efficiently with Snowpark, including setting up the environment, structuring the code, and utilizing it for different workloads.

In this chapter, we're going to cover the following main topics:

- Configuring the Snowpark development environment
- Operating with Snowpark
- Establishing a project structure for Snowpark

Technical requirements

For this chapter, you'll require an active Snowflake account and Python installed with Anaconda configured locally. You can refer to the following documentation for installation instructions:

- You can sign up for a Snowflake Trial account at `https://signup.snowflake.com/`
- To configure Anaconda, follow the guide at `https://conda.io/projects/conda/en/latest/user-guide/getting-started.html`.
- In addition, to install and set up Python for VS Code, follow the guide at `https://code.visualstudio.com/docs/python/python-tutorial`
- To learn how to operate a Jupyter Notebook in VS Code, go to `https://code.visualstudio.com/docs/datascience/jupyter-notebooks`

The supporting materials for this chapter are available in this book's GitHub repository at `https://github.com/PacktPublishing/The-Ultimate-Guide-To-Snowpark`.

Configuring the Snowpark development environment

The first step in developing Snowpark is to set up the Snowpark development environment. Developers have much flexibility regarding which **integrated development environments (IDEs)** they can use to get started with Snowpark; the only thing they need to do is install the Snowpark client **application programmable interface (API)** and connect to their Snowflake account. The development environment can be a local Python environment that contains your favorite IDE or the new Snowflake Python worksheets in Snowsight. This section will cover setting up a Snowpark Python worksheet and the local development environment.

Snowpark Python worksheet

Snowflake released *Snowflake Worksheets for Python*, a new type of worksheet in Snowsight for developing a Python-based Snowpark environment from within Snowflake. This game-changing feature allows developers to easily leverage the power of Snowpark Python within Snowsight to perform data processing and create data pipelines, **machine learning (ML)** models, and applications by integrating Snowpark Python directly into the browser without setting up Python environments or installing open source libraries on the client. Instead, developers can easily use pre-existing packages from Anaconda or import their own Python files from stages into the Worksheet. In addition, they can quickly deploy the Python worksheets as a stored procedure.

Prerequisites for using Python worksheets

To enable and use Snowflake Python worksheets, you must first acknowledge the Anaconda terms of service in Snowsight. The following steps are to be performed by the organization's administrator:

1. First, you must sign into your Snowflake account. In the **Switch Role** section on the left panel, switch to the **ORGADMIN** role in the user context:

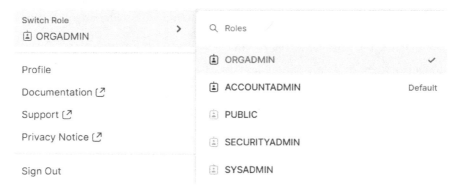

Figure 2.1 – The ORGADMIN role in Snowflake

2.   Then, go to **Admin | Billing & Terms**:

Figure 2.2 – Admin | Billing & Terms

3.   Click **Enable** next to **Anaconda Python packages**:

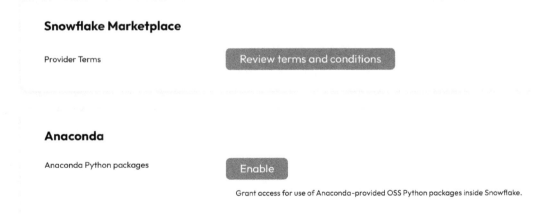

Figure 2.3 – Enabling Anaconda Python packages

4. You'll get a popup, as shown in the following screenshot. Click **Acknowledge & Continue** to enable the packages:

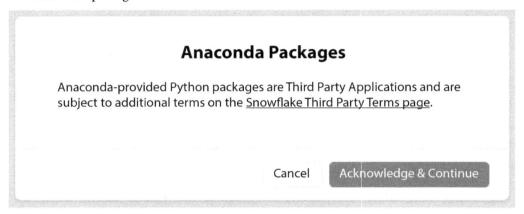

Figure 2.4 – Anaconda's Terms and Services

You'll need to enable the Anaconda packages every time you create a new Snowflake environment. Now that we've enabled the packages, let's see how we can work with Python worksheets.

Database and schema creation in Snowflake Snowsight

To create a database called SNOWPARK_DEFINITIVE_GUIDE and a schema called MY_SCHEMA using Snowflake's Snowsight UI interface, perform the following steps:

1. Go to the Snowsight web interface, log in using your Snowflake credentials, and navigate to the **Data | Databases** section. This is typically located in the left-hand navigation menu:

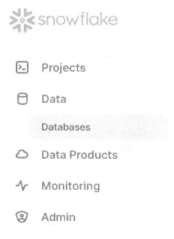

Figure 2.5 – The Databases section

2. Look for a button or link that says + **Database** in the top-right corner and click it. In the dialogue box that appears, enter SNOWPARK_DEFINITIVE_GUIDE as the name for the new database. Optionally, you can specify other settings, such as **Comment**, if needed, and click **Create**:

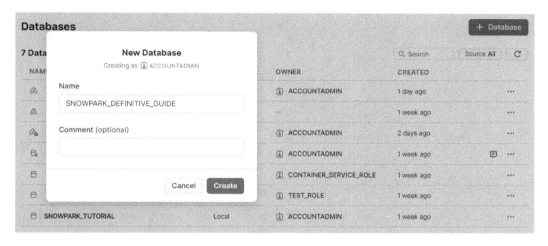

Figure 2.6 – The New Database dialogue

3. After creating the database, click on the SNOWPARK_DEFINITIVE_GUIDE database. This will take you to the following page:

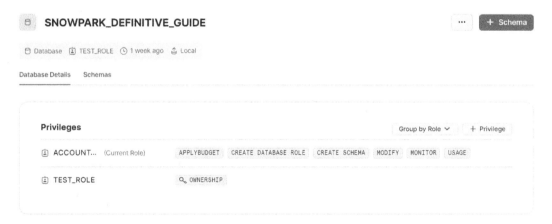

Figure 2.7 – The SNOWPARK_DEFINTIVE_GUIDE page

4. On this page, look for a button or link that says + **Schema** and click on it. In the dialogue box that appears, enter MY_SCHEMA as the name for the new schema. Optionally, you can specify other settings, such as **Comment** and **Managed access**. Click **Create** to create the schema:

Figure 2.8 – The New Schema dialogue

We'll consistently utilize the same database and schema throughout this book unless explicitly instructed otherwise in specific chapters.

Working with Python worksheets in Snowflake

Python worksheets come with a sample code template that can be used as a starter. The whole experience is embedded in Snowflake based on the Snowsight UI. Perform the following steps to create and work with Python worksheets in Snowflake:

1. Navigate to the **Worksheets** section from the menu in the Snowsight UI:

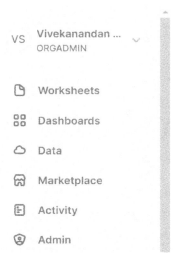

Figure 2.9 – The Worksheets menu option

2. On the **Worksheets** pane, click the + icon on the right and select **Python Worksheet** to create a Python worksheet:

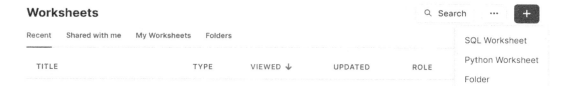

Figure 2.10 – Creating a Python worksheet

3. Select a database and schema so that you can work in the worksheet context. The scope of the Python code will operate based on these details:

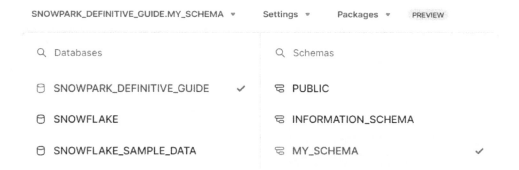

Figure 2.11 – Worksheet context

4. The worksheet has a handler function that's invoked when the worksheet is executed. The **Settings** menu allows you to configure your worksheets. The default handler function is main(), and the return type can be specified as Table(), Variant, or String. The result will be shown in the format you choose:

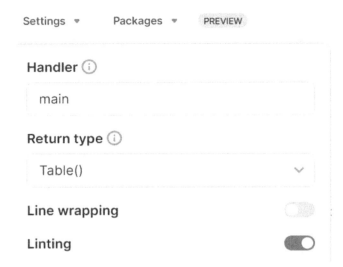

Figure 2.12 – Worksheet settings

Worksheet preferences such as linting and line wrapping can also be customized here.

5.   The Python code can be executed by clicking on the **RUN** button at the top right of the worksheet:

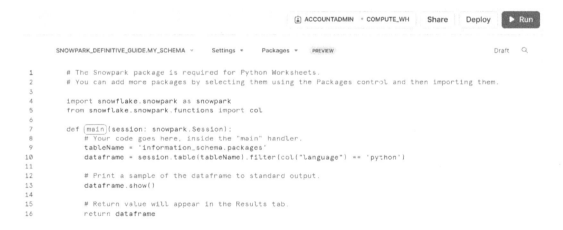

Figure 2.13 – Executing the worksheet

You can start developing and executing the Python code within the worksheet to see the results.

Managing Anaconda packages in Python worksheets

Python worksheets come with an integrated Anaconda environment that supports importing the most common Anaconda libraries without having to worry about managing dependencies. It also supports importing custom packages from the internal stage. In this section, we will look at managing Anaconda packages in Python. To do this, perform the following steps:

1. Navigate to the **Packages** tab from the menu and select **Anaconda Packages**. This will show a list of pre-installed packages and their versions:

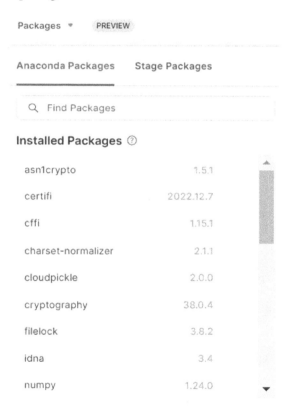

Figure 2.14 – Anaconda Packages

2. You can search for and install the required packages by using the search bar:

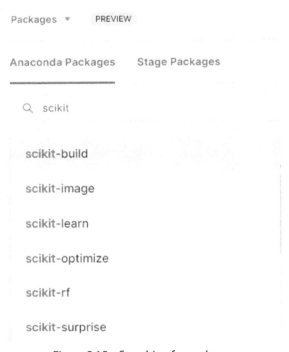

Figure 2.15 – Searching for packages

3. You can also modify the version of the packages by selecting the available versions from the respective dropdown:

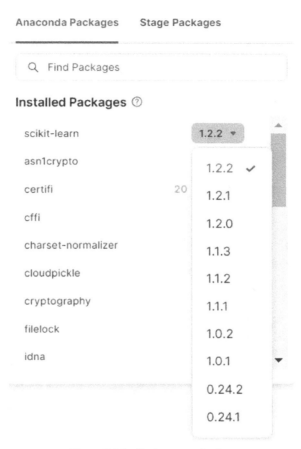

Figure 2.16 – Package versioning

4. You can execute the following query to check the available Python packages in Snowflake using SQL:

```
SELECT distinct package_name
FROM information_schema.packages
WHERE language = 'python';
```

This query obtains the results from the packages view in the information schema:

Figure 2.17 – Anaconda packages query

5. If you need to check the version information of specific packages, you can filter the query by the package's name:

```
SELECT *
FROM information_schema.packages
WHERE (package_name='numpy' AND language = 'python');
```

The output of the preceding code is as follows:

Figure 2.18 – Anaconda Python package query

This query shows the list of Anaconda packages that are available in my account. In the next section, we'll learn how to manage custom Python packages.

Managing custom packages in Python worksheets

Python worksheets also support the ability to import custom Python packages that can be used in Python code. However, the package must be uploaded into the internal stage, which is the storage part of the Snowflake account, and imported from it. In Snowsight, you can load files into a named internal stage area, allowing you to conveniently view and utilize them within your Python worksheets or load the data into a table using SQL. However, it's important to note that Snowsight does not support loading files into user or table stages.

To create a named internal stage, ensure that the role you are using has the **USAGE** privilege on the relevant database and schema and the **CREATE STAGE** privilege on the schema. Let's begin:

1. Sign in to Snowsight.

2. Access the **Data** section and navigate to **Databases**.

3. Choose the desired database and schema where you want to create the stage and load files.

4. Click **Create**, select **Stage**, and click **Snowflake Managed**:

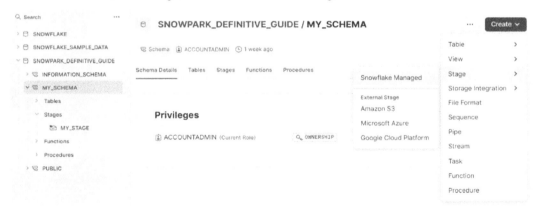

Figure 2.19 – Creating a stage

5.   Provide a name for the stage and opt-in to enable a directory table for the stage, allowing you to visualize the files. Once you're done, click **Create**:

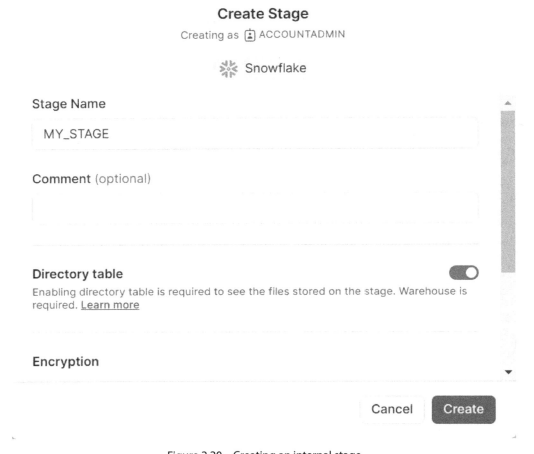

Figure 2.20 – Creating an internal stage

To load files into a Snowflake-managed named internal stage using Snowsight, follow these steps:

1.   Access the **Data** section and select **Databases**.

2.   Choose the database schema where you created the stage and select the stage itself.

3.   Click + **Files** to load the desired files into the stage.

4. In the **Upload Your Files** dialogue that appears, select the files you want to upload (multiple files can be selected simultaneously):

Figure 2.21 – Uploading a custom package

5. Optionally, specify or create a path within the stage where you want to store the files.

6. Click **Upload**.

7. Once the files have been successfully loaded into the stage, you can perform various actions, depending on your requirements.

With that, the package has been uploaded into the stage and is ready to be imported.

> **Important note**
>
> The maximum permissible file size is 50 MB. You need a role with the **USAGE** privilege on the database and schema and the **WRITE** privilege on the stage.

Once the stage has been created and the package has been uploaded, you can import the module so that you can use it in the program. To import a package, follow these steps:

1. From the **Packages** menu, select **Stage Packages**:

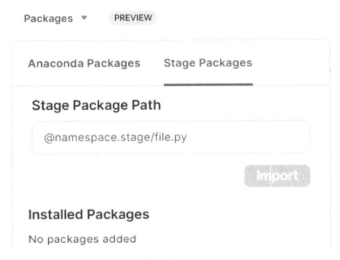

Figure 2.22 – Stage Packages

2. Enter the path to the package. You can refer to the stage in the same database and schema using `@Stage/path/to/package.py`. If the stage is in a different database and schema, you can use `@Database.Schema.Stage/path/to/package.py`:

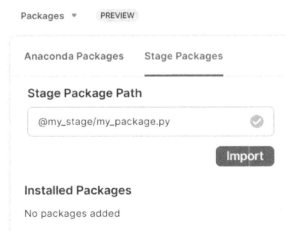

Figure 2.23 – Importing a stage package

3. Click **Import** to install the package. The module will now be visible under **Installed Packages**:

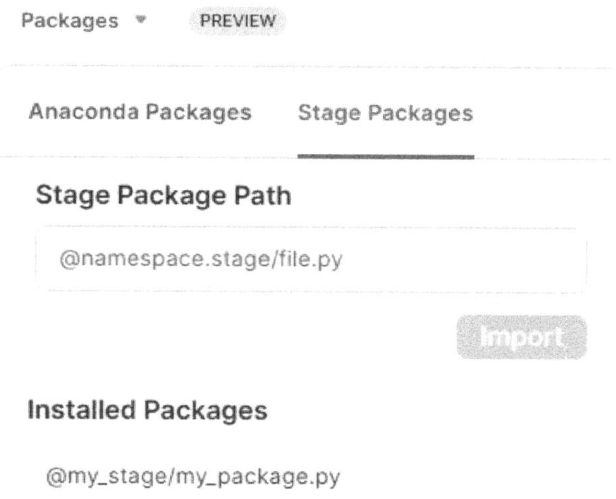

Packages ▼   PREVIEW

Anaconda Packages   Stage Packages

Stage Package Path

@namespace.stage/file.py

Import

Installed Packages

@my_stage/my_package.py

Figure 2.24 – Installing a stage package

4. You can import the package into your code by using import <package name>.

In the next section, we will cover how to deploy a Python stored procedure using the UI.

Deploying a Python stored procedure

The Python script in our worksheet can be seamlessly deployed as a stored procedure. This can then be used in a regular SQL context or scheduled to execute as a task. To deploy a stored procedure, follow these steps:

1. Click the **Deploy** button at the top right of the worksheet.

2. Give the stored procedure a name; an optional comment can also be specified that provides information about it. If the stored procedure already exists, then check the **Replace if exists** box to replace the existing stored procedure:

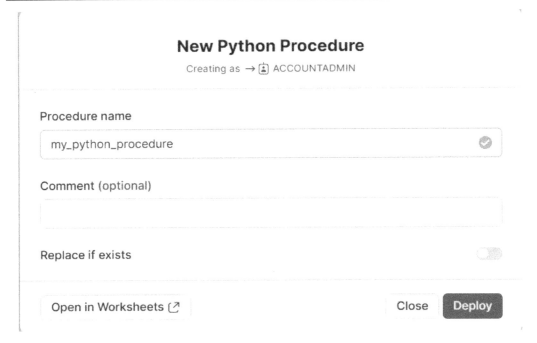

Figure 2.25 – Deploying a stored procedure

3. Click **Deploy** to deploy the stored procedure. It will be deployed under the database and schema you provided in the worksheet context.

4. The stored procedure can now be executed in your worksheets using the following code:

```
CALL DATABASE.SCHEMA.MY_STORED_PROCEDURE();
```

In the next section, we will cover the various features of Python worksheets.

Features of Python worksheets

Python worksheets consist of some features that are developer-friendly and support productivity. Here are some of the distinct characteristics of Python worksheets:

*   **Interactive Python environment**: Worksheets support the Python language and support Snowpark **user-defined functions** (**UDFs**) and stored procedures. Features such as syntax highlighting, type-sensitive autocomplete for keywords, and handy diagnostics such as undeclared variables or invalid method usage help increase developers' productivity:

```
import snowflake.snowpark as snowpark
from snowflake.snowpark.functions import col

def main(session: snowpark.Session):
    # Your code goes here, inside the "main" handler.
    session.ta
```

```
    table          def table(name: str | Iterable[str]) → Table
    method
                   Returns a Table that points the specified table.
    table_function
    method         Args:
                      name: A string or list of strings that specify the table
                   name or
    query_tag          fully-qualified object identifier (database name,
    property       schema name, and table name).

                      Note:
    use_database      If your table name contains special characters, use
    method         double quotes to mark it like this, session.table('"my
                   table"'). For fully qualified names, you need to use
                   double quotes separately like this, session.table('"my
                   db"."my schema"."my.table"').
                      Refer to Identifier Requirements.

                   Examples:
```

Figure 2.26 – Interactive Python environment

- **Python libraries support**: Python worksheets come with an integrated Anaconda environment that supports importing the most common Anaconda libraries without the need to worry about managing dependencies. It also supports importing custom packages from the internal stage.

- **Snowpark debugging**: Python worksheets can display the results from a DataFrame inside Snowsight using the show() or print() function. The preview of the DataFrame can also be returned to display the output in tabular format, which is very useful when it comes to debugging Python programs:

Figure 2.27 – Snowpark debugging

- **Collaboration**: Snowpark Python worksheets can be shared with developers. This makes collaboration much easier as multiple developers can access and work on the worksheet at the same time:

Figure 2.28 – Worksheet collaboration

- **Charts and data exploration**: Python worksheets provide a convenient way to visualize data and DataFrames as charts, which helps with data exploration and provides an easy way to analyze the data quickly:

Figure 2.29 – Charts and data exploration

In the next section, we will cover the limitations of Python worksheets.

Limitations of Python worksheets

Python worksheets are relatively new in Snowsight and have some limitations in terms of functionality. Let's take a closer look:

- Logging levels lower than **WARN** do not appear in the **Results** area by default.

- Python worksheets do not support breakpoints or selective execution of the portions of code. Instead, the entire code is run in the worksheet.

- The Python worksheet cannot display images or other artifacts of the Python code it generates. It can only show results that are returned through the DataFrame.

- It only supports Python 3.8 and libraries that use Python 3.8.

Snowflake constantly updates features for Python worksheets, improving them for developers. As a result, Python worksheets make it easier and faster for developers to code in Python from within the Snowflake environment.

Snowpark development in a local environment

Snowpark can also be developed conveniently from a local environment with your favorite IDE. The key benefit of Snowflake's partnership with Anaconda is Anaconda's Snowflake Snowpark for Python channel, which contains the necessary Python packages to run Snowpark. To use this channel, Anaconda or Miniconda should be installed and set up on the machine. In this section, we will walk you through how to set up Snowpark in your local development environment using Anaconda.

Important note

This book will utilize Anaconda for Snowpark development as it is the recommended approach and utilizes the benefits of Anaconda's package manager to easily set up and manage the Snowpark development environment.

The Snowpark API requires Python 3.8 to be installed. Snowflake recommends that you use Anaconda for easy package management. You can check out the Python version you have by running the following command in the **command-line interface (CLI)**:

```
python --version
```

Your output should look similar to the following:

```
(def_gui_3.8_env) C:\Users\Admin\Documents>python --version
Python 3.8.16
```

Figure 2.30 – Python CLI version

You can also check the Python version from within the Python code by running the following command:

```
from platform import python_version
print(python_version())
```

This will print an output similar to the following:

```
In [1]: from platform import python_version
        print(python_version())

        3.8.10
```

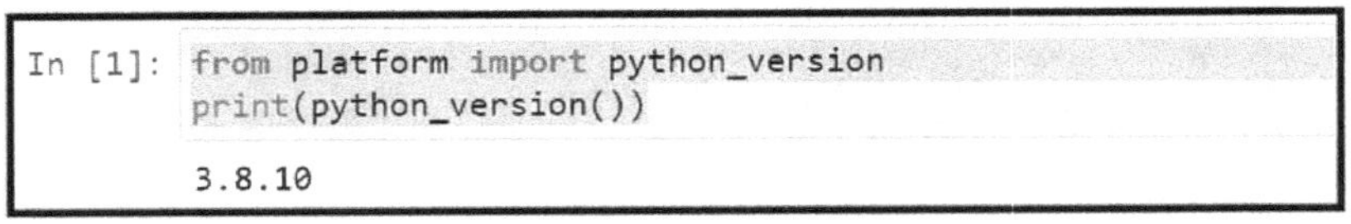

Figure 2.31 – Python version

Once Python has been installed, the virtual environment needs to be created. So, let's create one.

Creating a virtual environment

It's recommended that you create a Python virtual environment to ensure a seamless developer experience when working with Snowpark; it isolates the Snowpark API and allows you to manage all dependencies that are required for development. To create a virtual environment using Anaconda, run the following command:

```
conda create --name def_gui_3.8_env --override-channels --channel
https://repo.anaconda.com/pkgs/snowflake python=3.8
```

This command creates a new Python virtual environment named def_gui_3.8_env with Python 3.8 and installs the necessary packages, such as numpy and pandas, from Anaconda's Snowflake channel.

Installing the Snowpark Python package

Before installing the package, let's activate our Python virtual environment using the following command:

```
conda activate def_gui_3.8_env
```

The Snowpark package can be installed from Anaconda's Snowflake channel using the following command:

```
conda install --channel https://repo.anaconda.com/pkgs/snowflake
Snowflake-snowpark-python
```

Next, let's install some additional packages.

Installing additional Python packages

To install additional packages that are necessary for development, such as pandas and numpy, you can use the same Anaconda Snowflake channel:

```
conda install --channel https://repo.anaconda.com/pkgs/snowflake numpy
pandas
```

The virtual environment is now ready for development and is connected to your favorite IDE, Jupyter Notebook, or VS Code development. Similarly, we can install an IPython notebook.

Configuring Jupyter Notebook

Jupyter Notebook is one of the most popular IDEs for developers. In this section, we will cover how to configure the Jupyter IDE for Snowpark since the examples in this book use Jupyter. Jupyter Notebook needs to be installed in your local environment. The Jupyter environment comes installed alongside Anaconda. So, let's open Jupyter Notebook:

1. The def_gui_3.8_env virtual environment must be activated for development if it is not activated in the previous section. To activate the virtual environment, run the following command:

    ```
    conda activate def_gui_3.8_env
    ```

2. Launch Jupyter Notebook by running the following command:

    ```
    jupyter notebook
    ```

3. On the Jupyter Notebook web page, click the **New** button in the top-right corner. From the drop-down menu, select **Python 3** under the **Notebooks** section. This will open a new notebook with an empty cell that's ready for code execution.

Important note

This book will utilize Jupyter Notebook for all the examples.

Importing Snowpark modules

The Python classes for the Snowpark API are part of the snowflake.snowpark module. You can import particular classes from the module by specifying their names. For example, to import the average function, you can use the following code:

```
from snowflake.snowpark.functions import avg
```

Now that the development environment has been set up, let's learn how to operate with Snowpark.

Operating with Snowpark

Snowpark for Python consists of client APIs, UDFs, and stored procedures that execute directly on the Python engine. The following screenshot shows the various Snowpark objects that you can choose from:

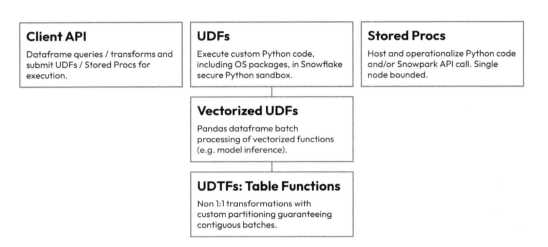

Figure 2.32 – Snowpark Python objects

Snowpark uses DataFrame objects to query and process data. The guiding principle in operating with Snowpark is to keep the data in Snowflake and process it right within Snowflake using the various Snowflake objects. The following figure shows Snowpark's architecture:

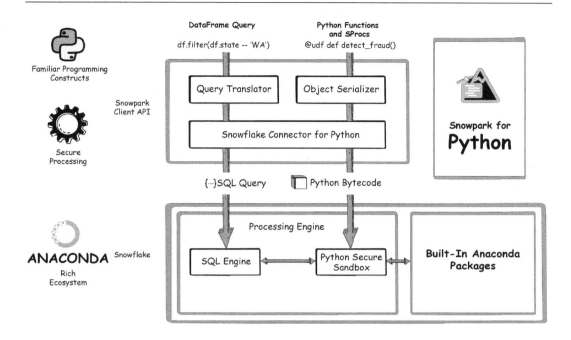

Figure 2.33 – Snowpark Python architecture

In the next section, we will cover the Python Engine.

The Python Engine

The Python Engine is an Anaconda-powered secure sandboxed Python environment that's executed on top of Snowflake's virtual warehouse and hosted on Snowflake's compute infrastructure. This lets you process data using Python without the need to extract the data outside the environment. The Python Engine consists of the UDF engine and the stored procedure engine. The UDF engine is a restricted engine that cannot read or write data outside of Snowflake, whereas the stored procedure engine is more permissive and consists of a session object for interacting with the Snowflake database.

Client APIs

The client API is the `snowflake-snowpark-python` library and it can be installed in any Python environment. It provides a session and the DataFrame APIs as methods to support queries being pushed down in Snowflake's Python Engine. The client APIs consist of notable objects, including sessions, DataFrames, and more. Let's look at these in detail.

Working with sessions

The Snowpark session is part of the Snowpark API and connects to Snowflake to interact with it and perform operations using Snowpark objects. The `Session` function in Snowpark API is responsible for operating with the session. To import the `Session` function for Snowpark, run the following command:

```
from snowflake.snowpark import Session
```

The Snowpark session consists of a Python dictionary containing the parameter values that are required to establish a connection to Snowflake:

```
connection_parameters = {
    "account": "<your snowflake account identifier>",
    "user": "<your snowflake username>",
    "password": "<your snowflake password>",
    "role": "<your snowflake role>", # optional
    "warehouse": "<your snowflake warehouse>",  # optional
    "database": "<your snowflake database>",  # optional
    "schema": "<your snowflake schema>" # optional
}
```

The connection consists of the following mandatory parameters:

- `Account`: The Snowflake account identifier is `<orgname>-<account_name>`. There is no need to specify the `snowflakecomputing.com` suffix.
- `User`: The username of the Snowflake user.
- `Password`: The password to authenticate to Snowflake.

The following are the optional parameters that can be passed to the connection:

- `role`: The role to be used in the Snowpark session. If left blank, the user's default role will be used.
- `warehouse`: The warehouse that's used to execute the process. If left blank, the user's default warehouse will be used.
- `database`: The database for the context of execution. If left blank, the user's default database will be used.
- `schema`: The schema for the context of execution. If left blank, the user's default schema will be used.

You can create the `session` object by passing this dictionary to the session builder, which is then used to establish the session:

```
session = Session.builder.configs(connection_parameters).create()
```

Once the session has been created, the `session` object acts as a handle to interact with Snowflake through various methods to perform operations such as reading and manipulating data. Some of the session methods that can be used are shown here:

```
print("Session Current Account:", session.get_current_account())
```

Finally, you can close the session by using the `close()` method. This terminates the session and the ongoing queries associated with it:

```
session.close()
```

Snowflake recommends that you close the session after the process has been completed. The `Session` method can be used to establish a session and interact with the `Session` objects.

Advanced authentication

Snowpark sessions also support advanced authentication, such as key pair authentication, if it's been configured for the user connecting to Snowflake. The private key must be serialized and then passed to the connection object of the session builder. We will be using the hazmat crypto package to serialize the private key. This private key is provided as a variable:

```
from cryptography.hazmat.backends import default_backend
from cryptography.hazmat.primitives import serialization
private_key_plain_text = '''-----BEGIN PRIVATE KEY-----
< your private key >
-----END PRIVATE KEY-----'''
private_key_passphrase = '<your private key passphase>'
```

We encode the private key using the passphrase and then serialize it to assign it to a variable that's passed into the Snowflake connection object:

```
private_key_encoded = private_key_plain_text.encode()
private_key_passphrase_encoded = private_key_passphrase.encode()
private_key_loaded = serialization.load_pem_private_key(
    private_key_encoded,
    password = private_key_passphrase_encoded,
    backend = default_backend()
)
private_key_serialized = private_key_loaded.private_bytes(
    encoding = serialization.Encoding.DER,
    format = serialization.PrivateFormat.PKCS8,
    encryption_algorithm = serialization.NoEncryption()
)
```

The connection parameters will have the private key passed in serialized form, and the session can be established using the session builder:

```
connection_parameters = {
    "account": "<your snowflake account identifier>",
    "user": "< your snowflake username>",
    "private_key": private_key_serialized,
    "warehouse": "<your snowflake warehouse>",
    "database": "<your snowflake database>",
    "schema": "<your snowflake schema>"
}
session = Session.builder.configs(connection_parameters).create()
```

Next, let's discuss Snowpark DataFrames.

Snowpark DataFrames

Snowpark for Python consists of a client API that provides a DataFrame-based approach that queries and processes data with a DataFrame object. The following diagram explains the code blocks that are utilized:

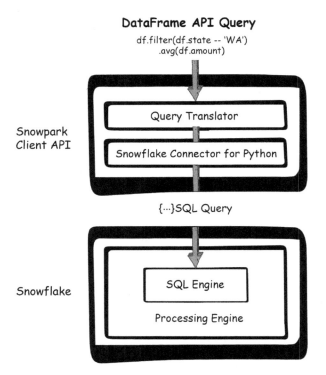

Figure 2.34 – Snowpark DataFrame API

The Snowpark DataFrame API yields more efficiency with less effort required by the developer as it has a concise syntax that is easy to understand and debug. The following figure compares a DataFrame and a SQL query:

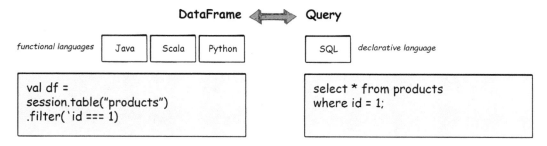

Figure 2.35 – Snowpark DataFrame versus a query

In addition, Snowpark DataFrames can directly read data from tables, views, and SELECT statements that support pushdown so that they can be executed in Snowflake. The Snowpark client APIs also support converting pandas DataFrames into Snowflake DataFrames so that data can be written back into Snowflake. Finally, Snowpark DataFrames support lazy evaluation as data computation is performed once an action is invoked.

> **Note**
>
> Lazy evaluation in Snowpark means that data processing operations are not executed immediately when they are defined. Instead, Snowpark builds a sequence of transformations without executing them until you explicitly request the result. This approach optimizes performance and resource usage, allowing you to construct complex data workflows efficiently and interactively. Lazy evaluation is a key feature for handling large datasets and optimizing data processing tasks in Snowpark.

Working with DataFrames

DataFrames are the dataset objects in Snowpark in which data is queried and processed. They represent relational datasets that provide lazy evaluation. The DataFrame executes SQL in a push-down manner and can perform operations such as creating objects and reading, writing, and working with data from the Python code. Various methods in the Session object are used to work with DataFrames.

Let's create an employee data table called SAMPLE_EMPLOYEE_DATA:

```
session.sql('CREATE OR REPLACE TABLE SAMPLE_EMPLOYEE_DATA (id INT,name
VARCHAR, age INT, email VARCHAR, city VARCHAR,country VARCHAR)').
collect()
```

The preceding code will create a table with the required fields for employee data. Let's insert some data into the table for operational purposes:

```
session.sql("""
    INSERT INTO SAMPLE_EMPLOYEE_DATA VALUES
    (1,'John Doe',25,'johndoe@example.com','New York','USA'),
    (2,'Jane Smith',30,'janesmith@example.com','Los Angeles','USA'),
    (3,'Michael Johnson',35,'michaeljohnson@example.com','London',
        'UK'),
    (4,'Sarah Williams',28,'sarahwilliams@example.com','Leeds',
        'UK'),
    (5,'David Brown',32,'davidbrown@example.com','Tokyo','Japan'),
    (6,'Emily Davis',29,'emilydavis@example.com','Sydney',
        'Australia'),
    (7,'James Miller',27,'jamesmiller@example.com','Dallas','USA'),
    (8,'Emma Wilson',33,'emmawilson@example.com','Berlin','Germany'),
    (9,'Alexander Taylor',31,'alexandertaylor@example.com',
        'Rome','Italy'),
    (10,'Olivia Anderson',26,'oliviaanderson@example.com',
        'Melbourne','Australia')
""").collect()
```

The preceding code will populate the table with the data that we can query. To query the data, we can directly pass the SQL statement:

```
session.sql("SELECT count(*) FROM SAMPLE_EMPLOYEE_DATA").collect()
```

The preceding code will return the results after it's been executed in Snowflake. We can also store these results in a DataFrame so that we can operate on it in Python:

```
from snowflake.snowpark.functions import col
df_subset_row = session.table(
    "SAMPLE_EMPLOYEE_DATA").filter(col("id") == 1)
```

The following code will save the results in a df_subset_row DataFrame that can be displayed using show():

```
df_subset_row.show()
```

Here's the output:

```
---------------------------------------------------------------------------------
|"ID"  |"NAME"    |"AGE"  |"EMAIL"                  |"CITY"     |"COUNTRY"  |
---------------------------------------------------------------------------------
|1     |John Doe  |25     |johndoe@example.com      |New York   |USA        |
---------------------------------------------------------------------------------
```

Figure 2.36 – DataFrame data

In the next section, we will look at Snowpark UDFs and stored procedures.

> **A note on code snippets**
>
> The examples presented in the following section have been simplified intentionally. Our main objective is to grasp and distinguish the concepts, rather than delving into complex scenarios. However, we'll delve into more sophisticated examples in the upcoming chapters.

UDFs

Snowpark for Python supports UDFs that allow developers to write reusable custom lambdas and functions to process the data through DataFrames. Like built-in functions, UDFs can be called from SQL, which enhances SQL with functionality that it doesn't have or doesn't do well. UDFs also provide a way to encapsulate functionality so that you can call it repeatedly from multiple places in your code. For example, you can write a UDF that returns a single value called a *scalar* function, also known as a UDF, or a group of values called a *tabular* function, also known as a **user data table function** (**UDTF**). These UDFs can be developed from within Python worksheets or by using Snowpark from your local development environment.

Scalar UDFs

Scalar UDFs are invoked once per row and return one output row for each input row. These UDFs are called just like a standard SQL function, with columns or expressions as arguments. It produces a row consisting of a single column/value as output. The data gets processed in parallel across each node within a multi-node virtual warehouse.

Working with UDFs

Once a Snowpark session has been created, the UDF can be turned into a standard function that can be registered in Snowflake:

```
def <main Python function name>(<arguments>):
    return <function output>
```

```
from snowflake.snowpark.types \
    import <specific Snowpark DataType object>

snowpark_session.add_packages(
    'List of native packages in Anaconda Channel')
snowpark_session.add_import('Path to Local File')

snowpark_session.udf.register(
    func = <Main Function Name>
  , return_type = <Return Type of Snowpark DataType object >
  , input_types = <[Input Types of Snowflake DataType object]>
  , is_permanent = True
  , name = '<UDF name>'
  , replace = True
  , stage_location = '@<UDF stage name>'
)
```

The preceding template delineates the steps in creating a UDF in Snowflake using Python. It involves defining the primary Python function that will be used by the UDF and registering it in Snowflake. Next, the function and UDF names are specified, along with the Snowflake stage, where the UDF files will be uploaded. Finally, the required Snowpark `DataType` object for the UDF's return value is imported, and its specific object is determined. This is not the only template we can follow – we can also leverage decorators to perform this. But for beginners, this can be very helpful to templatize and organize UDFS, UDTFs, and stored procedures.

Additionally, any necessary Snowpark `DataType` objects for the UDF's input arguments are imported and determined. The template also allows you to include extra packages and imports in the UDF. We can also specify whether the UDF should be temporary and whether an existing UDF with the same name should be overwritten:

```
def last_name_finder(input_name:str):
    last_name = input_name.split()[1]
    return last_name
from snowflake.snowpark.types \
    import StringType,IntegerType,ArrayType

test = session.udf.register(
    func = last_name_finder
  , return_type = StringType()
  , input_types = [StringType()]
  , is_permanent = True
  , name = 'LAST_NAME_FINDER'
  , replace = True
  , stage_location = '@MY_STAGE'
)
```

This simple function gets the input name and splits it so that it returns the last name. The UDF is registered to the internal My_Stage stage and is deployed into Snowflake. The UDF can be invoked directly in SQL as a function:

```
session.sql('''SELECT
    NAME,
    LAST_NAME_FINDER(NAME) AS LAST_NAME
    FROM SAMPLE_EMPLOYEE_DATA
''').show()
```

The output is as follows:

Figure 2.37 – UDF Snowpark execution

In this example, we invoked the LAST_NAME_FINDER function with the Name column, which returned the last name by splitting it. The function can also be called within the DataFrame function, as follows:

```
from snowflake.snowpark.functions import col, call_udf
df = session.table("SAMPLE_EMPLOYEE_DATA")
df.with_column(
    "last_name",call_udf("LAST_NAME_FINDER", col("name"))).show()
```

The preceding code generates the following output:

```
--------------------------------------------------------------------------------------------------------------
|"ID" |"NAME"           |"AGE" |"EMAIL"                        |"CITY"       |"COUNTRY"   |"LAST_NAME"  |
--------------------------------------------------------------------------------------------------------------
|1    |John Doe         |25    |johndoe@example.com            |New York     |USA         |Doe          |
|2    |Jane Smith       |30    |janesmith@example.com          |Los Angeles  |USA         |Smith        |
|3    |Michael Johnson  |35    |michaeljohnson@example.com     |London       |UK          |Johnson      |
|4    |Sarah Williams   |28    |sarahwilliams@example.com      |Leeds        |UK          |Williams     |
|5    |David Brown      |32    |davidbrown@example.com         |Tokyo        |Japan       |Brown        |
|6    |Emily Davis      |29    |emilydavis@example.com         |Sydney       |Australia   |Davis        |
|7    |James Miller     |27    |jamesmiller@example.com        |Dallas       |USA         |Miller       |
|8    |Emma Wilson      |33    |emmawilson@example.com         |Berlin       |Germany     |Wilson       |
|9    |Alexander Taylor |31    |alexandertaylor@example.com    |Rome         |Italy       |Taylor       |
|10   |Olivia Anderson  |26    |oliviaanderson@example.com     |Melbourne    |Australia   |Anderson     |
--------------------------------------------------------------------------------------------------------------
```

Figure 2.38 – UDF DataFrame execution

Next, let's look into UDTFs.

UDTF

Tabular UDFs, also known as UDTFs, require stateful operations to be performed on data batches and are invoked once per row, just like scalar UDFs, but they can return multiple rows as output for each input row. The UDTF handler method consists of an additional optional parameter that helps initialize the handler once for each partition and finalize processing for each section. A UDTF is a type of UDF that executes similarly to a UDF but with tabular output. Therefore, they can be developed in Python worksheets and Snowpark development environments.

Working with UDTFs

Creating a UDTF in Snowpark is similar to creating a UDF in that after a Snowpark session is created, the UDTF can be made directly in a standard command that can be registered in Snowflake.

The following Snowpark UDTF template provides a basic outline for creating a UDTF in Snowpark using Python. The following code shows the key elements in this template:

```
# Define Python class locally
'''
Define main Python class which is
leveraged to process partitions.
Executes in the following order:
- __init__  | Executes once per partition
- process | Executes once per input row within the partition
- end_partition | Executes once per partition
'''
class <name of main Python class> :
```

```
'''
Optional __init__ method to
execute logic for a partition
before breaking out into rows
'''
def __init__(self) :
```

This template creates a UDTF in Snowflake. First, a main Python handler class is defined for the UDTF, which can utilize other functions from the script or be imported from external sources. It is important to note that only one main Python handler class can be assigned to the UDTF.

In this template, you are expected to replace <name of main Python class> with a meaningful name for your UDTF class. This is the main class where you will define the logic for processing data within your UDTF.

The __init__ method is marked as optional, meaning you may or may not include it in your UDTF implementation. If you choose to include it, this method will execute once per partition before breaking out into individual rows.

You can use the __init__ method to perform any partition-level setup or initialization specific to your UDTF. For example, you might use it to initialize variables and open connections or set up data structures that will be used throughout the UDTF's execution:

```
'''
Method to process each input row
within a partition, returning a
tabular value as tuples.
'''
def process(self, <arguments>) :
    '''
    Enter Python code here that
    executes for each input row.
    This likely ends with a set of yield
    clauses that output tuples,
    for example:
    '''
    yield (<field_1_value_1>, <field_2_value_1>, ...)
    yield (<field_1_value_2>, <field_2_value_2>, ...)
    '''
    Alternatively, this may end with
    a single return clause containing
    an iterable of tuples, for example:
    '''
    return [
```

```
        (<field_1_value_1>, <field_2_value_1>, ...)
      , (<field_1_value_2>, <field_2_value_2>, ...)
    ]
```

This method is responsible for processing each input row within a partition and generating tabular data as tuples. Inside the `process` method, you can write custom Python code that executes for every input row in the partition. The key part of this method is the usage of `yield` statements or a `return` statement to produce tuples as output.

In terms of `yield` statements, you can output one or more tuples for each input row, allowing for flexibility in generating tabular results. Alternatively, you can use a `return` statement with a list of tuples to achieve the same result. In essence, the `process` method serves as the core logic for your UDTF, where you manipulate and transform data from each input row into tabular format, making it suitable for further processing or analysis:

```
'''
Optional end_partition method to
execute logic for a partition
after processing all input rows
'''
def end_partition(self) :
    # Python code at the partition level
    '''
    This ends with a set of yield
    clauses that output tuples,
    for example:
    '''
    yield (<field_1_value_1>, <field_2_value_1>, ...)
    yield (<field_1_value_2>, <field_2_value_2>, ...)
    '''
    Alternatively, this ends with
    a single return clause containing
    an iterable of tuples, for example:
    '''
    return [
        (<field_1_value_1>, <field_2_value_1>, ...)
      , (<field_1_value_2>, <field_2_value_2>, ...)
    ]
```

This method is used to execute logic that's specific to a partition after processing all the input rows in that partition. Inside the `end_partition` method, you can write custom Python code that performs calculations or generates results based on the data that's processed within that partition. This method can also be used to yield or return tabular data as tuples, similar to the `process` method, but this data typically represents aggregated or summarized information for the entire partition.

You have the option to use `yield` statements to output one or more tuples, or you can use a `return` statement with a list of tuples to provide the partition-level result. This allows you to perform partition-specific operations and return the results in a structured format.

The `end_partition` method in a Snowpark UDTF template is used for executing partition-level logic and returning tabular data or results specific to that partition after processing all input rows within it. It's especially useful for tasks such as aggregations or calculations, which require data from the entire partition.

The following code template provides details on how to register a UDTF in Snowflake and the corresponding options to define the UDTF:

```
from snowflake.snowpark.types import StructType, StructField

from snowflake.snowpark.types \
    import <specific Snowpark DataType objects>

snowpark_session.add_packages(
    '<list of required packages natively available in Snowflake(
        i.e. included in Anaconda Snowpark channel)>')
snowpark_session.add_import('<path\\to\\local\\directory\\or\\file>')

snowpark_session.udtf.register(
    handler = <name of main Python class>
  , output_schema = StructType(
        <list of StructField objects with specific field \
        name and Snowpark DataType objects>)
  , input_types = <list of input DataType() \
        objects for input parameters>
  , is_permanent = True
  , name = '<UDTF name>'
  , replace = True
  , stage_location = '@<UDTF stage name>'
  )
```

The `udtf()` method of the Snowflake Snowpark `Session` object is used to create the UDTF in Snowflake. The process involves several steps: determining the Python class that the UDTF will use, specifying the name of the UDTF within Snowflake (it can be a fully qualified name or created in the same namespace as the Snowpark `Session` object), and providing the name of the Snowflake stage where the UDTF files will be uploaded.

Specific Snowpark `DataType` objects are imported to define the structure of the UDTF. This includes importing objects for defining tabular structures, such as table schemas (using `StructType`) and fields within a table (using `StructField`). Furthermore, a specific Snowpark `DataType` object is imported for the values that are passed into and returned by the UDTF. The output schema of the UDTF is defined using the imported Snowpark `DataType` objects, embedding them into `StructField` and `StructType` objects. Additionally, a list of specific Snowpark `DataType` objects is defined for the input arguments of the UDTF. It is crucial to include all these `DataType` objects in the import and ensure they match the expected arguments that are passed to the `process` method within the `handler` class.

The template allows a temporary UDTF to be created that only exists within the specific Snowflake Snowpark `Session` object. Additionally, an option exists to overwrite an existing UDTF with the same name; an error will be returned if it's set to `False` and a UDTF already exists. Lastly, the template briefly mentions adding additional packages and imports to the UDTF, which is optional and can be done using the provided rows.

The following example illustrates how to use a Snowpark UDTF to calculate the averages of numeric data within Snowflake tables. This showcases the practical application of UDTFs in Snowpark for custom data processing tasks:

```
class CalculateAverage:
    def __init__(self) :
        self._values = []

    def process(self, input_measure: int) :
        self._values.append(input_measure)

    def end_partition(self) :
        values_list = self._values
        average = sum(values_list) / len(values_list)
        yield(average ,)
```

The `CalculateAverage` Snowpark UDTF is designed to compute the average of a numeric column within a Snowflake table. It does this by accumulating the input values for each partition of the data and then calculating the average when the partition ends.

The `process` method collects input values one by one and stores them in a list. When the partition ends (in the `end_partition` method), it calculates the average by summing up all the collected values and dividing by the count of values. Finally, it yields the calculated average as the UDTF's output. This UDTF simplifies the process of computing averages in Snowflake SQL queries, especially when dealing with large datasets:

```
from snowflake.snowpark.types import StructType, StructField
from snowflake.snowpark.types \
    import FloatType,IntegerType,StringType

output_schema = StructType([
    StructField("Avg_Age", FloatType())
])
```

```
session.udtf.register(
    handler = CalculateAverage
  , output_schema = output_schema
  , input_types = [IntegerType()]
  , is_permanent = True
  , name = 'AVERAGE_AGE'
  , replace = True
  , stage_location = '@MY_STAGE'
)
```

In this example, we're creating a UDTF function called Average_Age that calculates the average age by getting the age as input. The function is uploaded into MY_STAGE and registered in Snowflake.

The function can be executed to get the average age per country from the sample employee data:

```
session.sql('''
    SELECT
        COUNTRY, Avg_Age
    FROM
        SAMPLE_EMPLOYEE_DATA,
        table(AVERAGE_AGE(AGE) over (partition by COUNTRY))
''').show()
```

This will display the following output:

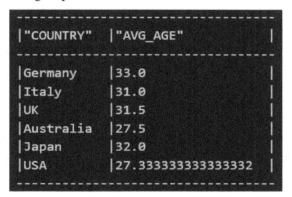

Figure 2.39 – UDTF Snowpark execution

The output shows the execution output of the UDTF. In the next section, we will cover vectorized UDFs.

Vectorized UDFs

Vectorized UDFs operate similarly to scalar UDFs in that they let you define Python functions that receive batches of input rows as pandas DataFrames and return collections of results as pandas arrays or series.

Vectorized UDFs parallelize the operation on batches of data and provide significant performance advantages on sets of rows compared to serial row processing. In addition, they reduce the complexity of using libraries that operate on pandas DataFrames and arrays.

Working with vectorized UDFs

The same example we looked at previously can be executed in the Snowpark environment after establishing the session by passing a vectorized DataFrame as the input to the standard UDF:

```python
import pandas as pd
from snowflake.snowpark.functions import pandas_udf
from snowflake.snowpark.types \
    import IntegerType, PandasSeriesType,StringType

@pandas_udf(
    name='column_adder'
  , stage_location = '@MY_STAGE'
  , input_types=[PandasSeriesType(StringType()), \
        PandasSeriesType(StringType())]
  , return_type=PandasSeriesType(StringType())
  , is_permanent=True
  , replace=True)
def column_adder(
    column1: pd.Series, column2: pd.Series) -> pd.Series:
    return column1 + "," + column2

df = session.table("SAMPLE_EMPLOYEE_DATA")
df.withColumn('City_Country', column_adder(col('CITY'), \
    col('COUNTRY'))).show()
```

The example UDF returns the city and country in the CITY_COUNTRY column for each row in the Sample Employee table:

```
|"ID" |"NAME"           |"AGE" |"EMAIL"                       |"CITY"     |"COUNTRY"  |"CITY_COUNTRY"       |
-------------------------------------------------------------------------------------------------------------
|1    |John Doe         |25    |johndoe@example.com           |New York   |USA        |New York,USA         |
|2    |Jane Smith       |30    |janesmith@example.com         |Los Angeles|USA        |Los Angeles,USA      |
|3    |Michael Johnson  |35    |michaeljohnson@example.com    |London     |UK         |London,UK            |
|4    |Sarah Williams   |28    |sarahwilliams@example.com     |Leeds      |UK         |Leeds,UK             |
|5    |David Brown      |32    |davidbrown@example.com        |Tokyo      |Japan      |Tokyo,Japan          |
|6    |Emily Davis      |29    |emilydavis@example.com        |Sydney     |Australia  |Sydney,Australia     |
|7    |James Miller     |27    |jamesmiller@example.com       |Dallas     |USA        |Dallas,USA           |
|8    |Emma Wilson      |33    |emmawilson@example.com        |Berlin     |Germany    |Berlin,Germany       |
|9    |Alexander Taylor |31    |alexandertaylor@example.com   |Rome       |Italy      |Rome,Italy           |
|10   |Olivia Anderson  |26    |oliviaanderson@example.com    |Melbourne  |Australia  |Melbourne,Australia  |
-------------------------------------------------------------------------------------------------------------
```

Figure 2.40 – Vectorized UDF in Snowpark

The output shows the execution output of the vectorized UDF. In the next section, we will cover stored procedures.

Stored procedures

A Python stored procedure is a series of code statements you can parameterize and execute on demand. They run in a less restricted environment than UDFs and support interacting with Snowflake objects, as well as performing DDL and DML operations on tables.

Stored procedures in Snowpark are utilized for executing tasks and streams within Snowflake's data processing framework. These stored procedures encapsulate specific logic or functionality, allowing users to perform various operations on data seamlessly. Tasks, which are often associated with batch processing, involve executing predefined actions or workflows on datasets at scheduled intervals. Stored procedures enable users to automate these tasks, ensuring consistent and efficient data processing.

On the other hand, streams are continuous data pipelines that capture changes in real-time from a data source. Stored procedures play a vital role in managing and processing streams by defining how incoming data should be processed and integrated into the target destination. With Snowpark, users can create stored procedures to handle these streaming data scenarios, including data transformation, filtering, and loading data into Snowflake tables.

Working with stored procedures

A stored procedure can be created in Snowpark with the following template:

```
# Define Python function locally
def <Python Function Name>(
    snowpark_session: snowflake.snowpark.Session, <arguments>):
    return <Output>
```

```
# Imports Required For Stored Procedure
from snowflake.snowpark.types \
    import <specific Snowpark DataType object>

# Optional: Import additional packages or files
snowpark_session.add_packages(
    'List of native packages in Anaconda Channel')
snowpark_session.add_import('Path to Local File')

# Upload Stored Procedure to Snowflake
snowpark_session.sproc.register(
    func = <Function name to register>
  , return_type = <Return Type of Snowpark DataType object>
  , input_types = <[Input Types of Snowflake DataType object]>
  , is_permanent = True
  , name = '<Stored Procedure name>'
  , replace = True
  , stage_location = '@<Stored Procedure stage name>'
    <optional: , execute_as = 'CALLER'>
)
```

Here, we define the main Python function that will be used in the stored procedure and an additional argument called snowpark_session, which allows us to interact with Snowflake objects. Next, we use the sproc.register() method to create the stored procedure, specifying the Python function, stored procedure's name, and Snowflake stage for file uploads. Finally, we import specific Snowpark DataType objects for the stored procedure's return value and input arguments.

The snowflake_session argument is implicitly understood and not included in the input arguments. Optional rows allow for additional packages and imports. Here, we can determine whether the stored procedure will be temporary. We can also decide whether to overwrite an existing one with the same name and specify whether it will execute as the caller or the owner:

```
def subset_table(snowpark_session:Session):
    df = snowpark_session.table(
        'SAMPLE_EMPLOYEE_DATA').select("NAME","AGE")
    return df.collect()

from snowflake.snowpark.types import StringType
session.add_packages('snowflake-snowpark-python')

session.sproc.register(
    func = subset_table
  , return_type = StringType()
  , input_types = []
```

```
    , is_permanent = True
    , name = 'SPROC_SUBSET_TABLE'
    , replace = True
    , stage_location = '@MY_STAGE'
  )
```

The stored procedure returns the column name and the age from the `Employee Data` table. It's registered as `SPROC_SUBSET_TABLE` and uploaded through `My_Stage`:

```
session.sql(''' CALL SPROC_SUBSET_TABLE()''').show()
```

Here's the output:

```
--------------------------------------------------------------
|"SPROC_SUBSET_TABLE"                                        |
--------------------------------------------------------------
|[Row(NAME='John Doe', AGE=25), Row(NAME='Jane S...        |
--------------------------------------------------------------
```

Figure 2.41 – Stored procedure execution

The stored procedure can be executed by running the `CALL` command.

The difference between UDFs and stored procedures

UDFs and stored procedures have significant differences in terms of functionality and usage. The following figure shows the basic differences between UDFs and stored procedures and what they're used for:

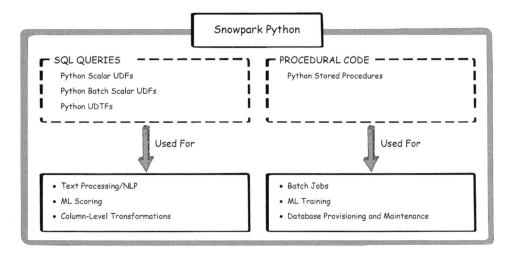

Figure 2.42 – UDFs versus stored procedures

The following table shows the differences and similarities between UDFs and stored procedures based on their properties:

	UDFs	Stored Procedures
Purpose	Perform calculations and return the results. UDFs require a value to be returned.	Perform complex operations by executing SQL statements. They do not require an explicit value to be returned.
Usage	UDFs can be used when logic needs to be called as part of SQL statements that return a value.	When database operations or administrative tasks need to be performed.
Output	UDFs always need to return a result.	Stored procedures don't need to return a result.
Context	UDF return values are directly accessible in the SQL context.	Stored procedure return values are not accessible in the SQL context.
Execution	UDFs can be called in the context of another SQL statement. In addition, multiple UDFs can be invoked in a single SQL statement.	Stored procedures are called independently. Therefore, only a single stored procedure is invoked in a SQL statement.
Security	UDFs cannot access the database or perform operations directly on it.	Stored procedures can access the database and perform data operations on it.

Table 2.1 – Comparison of UDFs and stored procedures

Both stored procedures and UDFs can be used together to expand the capabilities of Python execution in Snowflake.

Establishing a project structure for Snowpark

To assist with the development of Snowpark in Python and to make it easy to create a Snowpark project, Snowflake has released Snowpark project templates for Python. These contain everything you'll need for developing, testing, and deploying with Snowpark – they provide all the boilerplate required to develop UDFs and stored procedures, along with unit tests and even GitHub Actions workflow files for CI/CD.

The project template has been released as open source on GitHub, making it easy for developers to clone and use the project. To clone the project, follow these steps:

1. Download the files or clone the repository from `https://github.com/Snowflake-Labs/snowpark-python-template`. A new GitHub repository can be created from the template by using the GitHub CLI, like so:

    ```
    gh repo create <new-repo-name> --template="Snowflake-Labs/
    snowpark-python-template"
    ```

    The new repository's name needs to be specified. The repository will be similar to the Snowpark project template.

2. Set up the following environment variables so that you can configure the necessary Snowflake details:

    ```
    SNOWSQL_ACCOUNT=<replace with your account identifier>
    SNOWSQL_USER=<replace with your username>
    SNOWSQL_PWD=<replace with your password>
    SNOWSQL_DATABASE=<replace with your database>
    SNOWSQL_SCHEMA=<replace with your schema>
    SNOWSQL_WAREHOUSE=<replace with your warehouse>
    ```

    These environment variables are required to connect to the Snowflake environment and for Snowpark to establish a session.

3. Create an Anaconda virtual environment and install the dependencies from the `environment.yml` file:

    ```
    conda env create -f environment.yml
    conda activate snowpark
    ```

    The preceding code creates an Anaconda environment called `snowpark` that can be used for development.

4. You can test the connection and check if the environment has been set up by executing the `app.py` stored procedure. Navigate to the project folder and run the following command:

    ```
    python src/procs/app.py
    ```

    This produces an output called **Hello World** that establishes a connection to Snowflake.

The Snowpark project supports functions and stored procedures. The project structure consists of the `procs` directory for stored procedures, the `udf` directory for UDFs, and the `util` directory for the utilities methods and classes that are shared between UDFs and stored procedures.

The `test` folder consists of the test cases that can be tested via `pytest`. There's also a GitHub workflow that can be deployed via GitHub Actions. We will cover this in detail in the following chapters.

Summary

Snowpark is very versatile and supports complex development patterns. In this chapter, we learned how to configure the Snowpark development environment and different Snowpark objects, such as sessions, UDFs, and stored procedures, and how to use them. We also learned how to set up a Snowpark development project locally and inside a Python worksheet before looking at some sample code that we could use to start developing.

In the next chapter, we will cover how to perform data processing with Snowpark, as well as how to ingest, prepare, and analyze data.

Part 2: Snowpark Data Workloads

This part focuses on deploying data workloads such as data engineering, data science, and **machine learning (ML)** models in Snowpark.

This part includes the following chapters:

- *Chapter 3, Simplifying Data Processing Using Snowpark*
- *Chapter 4, Building Data Engineering Pipelines with Snowpark*
- *Chapter 5, Developing Data Science Projects with Snowpark*
- *Chapter 6, Deploying and Managing ML Models with Snowpark*

3

Simplifying Data Processing Using Snowpark

In the previous chapter, we learned how to set up a development environment for Snowpark, as well as various Snowpark components, such as DataFrames, UDFs, and stored procedures. We also covered how to operate those objects and run them in Snowflake. In this chapter, we will cover data processing with Snowpark and learn how to load, prepare, analyze, and transform data using Snowpark.

In this chapter, we're going to cover the following main topics:

- Data ingestion
- Data exploration and transformation
- Data grouping and analysis

Technical requirements

For this chapter, you require an active Snowflake account and Python installed with Anaconda configured locally. You can refer to the following documentation for installation instructions:

- You can sign up for a Snowflake Trial account at `https://signup.snowflake.com/`
- To configure Anaconda, follow the guide at `https://conda.io/projects/conda/en/latest/user-guide/getting-started.html`
- To install and set up Python for VS Code, follow the guide at `https://code.visualstudio.com/docs/python/python-tutorial`
- To learn how to operate Jupyter Notebook in VS Code, go to `https://code.visualstudio.com/docs/datascience/jupyter-notebooks`

The supporting materials for this chapter are available in this book's GitHub repository at `https://github.com/PacktPublishing/The-Ultimate-Guide-To-Snowpark`.

Data ingestion

The first part of the data engineering process is data ingestion – it is crucial to get all the different data into a usable format in Snowflake for analytics. In the previous chapter, we learned how Snowpark can access data through a DataFrame. This DataFrame can access data from Snowflake tables, views, and objects, such as streams, if we run a query against it. Snowpark supports structured data in various formats, such as Excel and CSV, as well as semi-structured data, such as JSON, XML, Parquet, Avro, and ORC; specialized formats, such as HL7 and DICOM, and unstructured data, such as images and media, can be ingested and handled in Snowpark. Snowpark enables secure and programmatic access to files in Snowflake stages.

The flexibility of Snowpark Python allows you to adapt to changing data requirements effortlessly. Suppose you start with a CSV file as your data source; you can switch to a JSON or packet format at a later stage. With Snowpark, you don't need to rewrite your entire code base. Instead, you can make minor adjustments or configuration changes to accommodate the new structure while keeping the core logic intact. This flexibility saves you valuable time and effort, enabling you to switch between different data formats as your needs evolve quickly.

By leveraging Snowpark's capabilities, you can focus more on analyzing and utilizing data rather than worrying about the intricacies of data format handling. This streamlined approach empowers you to experiment with different data sources, adapt to evolving data requirements, and efficiently load data into Snowflake tables, all with minimal code changes and maximum flexibility.

So, let's delve into the power of Snowpark Python and its ability to effortlessly handle different data formats, allowing you to work with diverse sources without cumbersome code modifications. You will experience the freedom to explore, analyze, and extract insights from your data while enjoying a seamless and flexible integration process.

The data ingestion scripts are provided in this book's GitHub repository: `https://github.com/ PacktPublishing/The-Ultimate-Guide-To-Snowpark`. These scripts will simplify the process of uploading any new dataset that will be used for analysis, ensuring a smooth and efficient workflow. Following a similar approach to what was outlined in the preceding chapters, you can effortlessly upload new datasets and explore Snowflake's data engineering and machine learning functionalities. The provided data ingestion scripts will act as your guide, making the process seamless and hassle-free.

Important note on datasets

The dataset we'll be using in this chapter provides unique insights into customer behavior, campaign responses, and complaints, enabling data-driven decision-making and customer satisfaction improvement. The original dataset is from the Kaggle platform (`https://www.kaggle.com/datasets/ rodsaldanha/arketing-campaign`). However, the datasets that will be discussed in this section are not directly accessible via a Kaggle link. Instead, we started with a base dataset and generated new data formats to illustrate loading various dataset formats using Snowpark. These datasets can be found in this book's GitHub repository under the `datasets` folder.

The datasets include purchase history in CSV format, campaign information in JSON format, and complaint information in Parquet format. These datasets provide valuable information about customer behavior, campaign responses, and complaints:

- **Purchase history (CSV)**: This file contains customer information, such as ID, education, marital status, and purchase metrics. The dataset offers insights into customer buying habits and can be further analyzed for data-driven decisions.

- **Campaign information (JSON)**: The JSON dataset includes data on campaign acceptance and customer responses. Analyzing this dataset will help you refine marketing strategies and understand campaign effectiveness.

- **Complaint information (Parquet)**: This file contains details about customer complaints, including contact and revenue metrics. This dataset aids in tracking and addressing customer complaints for improved satisfaction.

> **Note**
>
> Moving forward, we will be utilizing our local development environment to execute all Snowpark code, rather than relying on Snowflake worksheets. This approach offers greater flexibility and control over the development and testing of Snowpark scripts. When worksheets are used for specific tasks, we will explicitly call out their usage for clarity and context.

Ingesting a CSV file into Snowflake

Snowflake supports ingesting data easily using CSV files. We will load the purchase history data into the PURCHASE_HISTORY table as a CSV file. We'll upload purchase_history.csv into an internal stage by using a Snowpark session, as shown here:

```
session.file.put('./datasets/purchase_history.csv', 'MY_STAGE')
```

With that, the file has been uploaded to the internal stage. We will reference this directly in Snowpark. The data schema for the marketing table can also be directly defined as a Snowpark type. The following code provides the necessary columns and data types to create the table in Snowflake:

```
import snowflake.snowpark.types as T
purchase_history_schema = T.StructType([
    T.StructField("ID", T.IntegerType()),
    T.StructField("Year_Birth", T.IntegerType()),
    T.StructField("Education", T.StringType()),
    T.StructField("Marital_Status", T.StringType()),
    T.StructField("Income", T.IntegerType()),
    T.StructField("Kidhome", T.IntegerType()),
    T.StructField("Teenhome", T.IntegerType()),
    T.StructField("Dt_Customer", T.DateType()),
```

```
    T.StructField("Recency", T.IntegerType()),
    T.StructField("MntWines", T.IntegerType()),
    T.StructField("MntFruits", T.IntegerType()),
    T.StructField("MntMeatProducts", T.IntegerType()),
    T.StructField("MntFishProducts", T.IntegerType()),
    T.StructField("MntSweetProducts", T.IntegerType()),
    T.StructField("MntGoldProds", T.IntegerType()),
    T.StructField("NumDealsPurchases", T.IntegerType()),
    T.StructField("NumWebPurchases", T.IntegerType()),
    T.StructField("NumCatalogPurchases", T.IntegerType()),
    T.StructField("NumStorePurchases", T.IntegerType()),
    T.StructField("NumWebVisitsMonth", T.IntegerType())
])
```

In this code snippet, we take our first step toward understanding the structure of our data by defining a schema for our purchase history dataset. Using the Snowflake Snowpark library, we establish the fields and corresponding data types, setting the foundation for our data analysis journey. This code serves as a starting point, guiding us in defining and working with structured data. This is not the only way we can load the dataset using Snowpark. We will continue to explore different methodologies to load other tabular datasets as we progress.

This code imports the necessary types from the Snowflake Snowpark library. It creates a variable called purchase_history_schema and assigns it a StructType object, representing a structured schema for the dataset. The StructType object contains multiple StructField objects, each representing a field in the dataset. Each StructField object specifies the name of the area and its corresponding data type using the types provided by Snowflake Snowpark. The following code reads the file:

```
purchase_history = session.read\
        .option("FIELD_DELIMITER", ',')\
        .option("SKIP_HEADER", 1)\
        .option("ON_ERROR", "CONTINUE")\
        .schema(purchase_history_schema).csv(
            "@MY_Stage/purchase_history.csv.gz")\
        .copy_into_table("PURCHASE_HISTORY")
```

The CSV file is read with file format options such as FIELD_DELIMITER, SKIP_HEADER, and others, all of which are specified alongside the schema defined in the preceding definition. The PURCHASE_ HISTORY table was created with the data from the CSV file, which is now ready for processing:

```
session.table("PURCHASE_HISTORY").show()
```

The preceding code shows the output of the PURCHASE_HISTORY table:

"ID"	"YEAR_BIRTH"	"EDUCATION"	"MARITAL_STATUS"	"INCOME"	"KIDHOME"	"TEENHOME"	"DT_CUSTOMER"	"RECENCY"
5524	1957	Graduation	Single	58138	0	0	2012-09-04	58
2174	1954	Graduation	Single	46344	1	1	2014-03-08	38
4141	1965	Graduation	Together	71613	0	0	2013-08-21	26
6182	1984	Graduation	Together	26646	1	0	2014-02-10	26
5324	1981	PhD	Married	58293	1	0	2014-01-19	94
7446	1967	Master	Together	62513	0	1	2013-09-09	16
965	1971	Graduation	Divorced	55635	0	1	2012-11-13	34
6177	1985	PhD	Married	33454	1	0	2013-05-08	32
4855	1974	PhD	Together	30351	1	0	2013-06-06	19
5899	1950	PhD	Together	5648	1	1	2014-03-13	68

Figure 3.1 – The PURCHASE_HISTORY table

The CSV is easy to load as it uses the file format options available in Snowflake. Now, let's see how we can load JSON files into Snowflake.

Ingesting JSON into Snowflake

Snowflake allows JSON structures to be ingested and processed via the Variant data type. We can ingest JSON similar to how we would ingest a CSV file – by uploading it into the internal stage. The campaign_info.json file contains data about marketing campaigns. We can load this into the CAMPAIGN_INFO table by using the following code:

```
session.file.put('./datasets/campaign_info.json', 'MY_STAGE')
```

With that, the file has been uploaded to the internal stage; we will reference it in Snowpark. Snowpark can access the file to load it into a table:

```
df_from_json = session.read.json("@My_Stage/campaign_info.json.gz")
```

The contents of the JSON file are read into the DataFrame as JSON objects. This DataFrame can be written into a table as a variant:

```
df_from_json.write.save_as_table("CAMPAIGN_INFO_TEMP",
    mode = "overwrite")
```

The CAMPAIGN_INFO_TEMP table contains the JSON data. We can query the table to view the data:

```
df_from_json.show()
```

The preceding command displays the JSON data from the DataFrame:

```
-------------------------------
|"$1"                         |
-------------------------------
|{                            |
|    "AcceptedCmp1": 0,       |
|    "AcceptedCmp2": 0,       |
|    "AcceptedCmp3": 0,       |
|    "AcceptedCmp4": 0,       |
|    "AcceptedCmp5": 0,       |
|    "ID": 5524,              |
|    "Response": 1            |
|}                            |
|{                            |
|    "AcceptedCmp1": 0,       |
|    "AcceptedCmp2": 0,       |
|    "AcceptedCmp3": 0,       |
|    "AcceptedCmp4": 0,       |
|    "AcceptedCmp5": 0,       |
|    "ID": 2174,              |
|    "Response": 0            |
|}                            |
```

Figure 3.2 – The Campaign Info table

The following code snippet utilizes the Snowpark library in Snowflake to manipulate a DataFrame:

```
from snowflake.snowpark.functions import col
df_flatten = df_from_json.select(col("$1")["ID"].as_("ID"),\
    col("$1")["AcceptedCmp1"].as_("AcceptedCmp1"),\
    col("$1")["AcceptedCmp2"].as_("AcceptedCmp2"),\
    col("$1")["AcceptedCmp3"].as_("AcceptedCmp3"),\
    col("$1")["AcceptedCmp4"].as_("AcceptedCmp4"),\
    col("$1")["AcceptedCmp5"].as_("AcceptedCmp5"),\
    col("$1")["Response"].as_("Response"))
df_flatten.write.save_as_table("CAMPAIGN_INFO")
```

The preceding code selects specific columns from an existing DataFrame and renames them using the col function. The transformed DataFrame is then saved as a new table in Snowflake. The code performs data **extraction, transformation, and loading (ETL)** operations by selecting and renaming columns within the DataFrame and saving the result as a new table in Snowflake.

The CAMPAIGN_INFO table now contains the flattened data, with the data in separate columns so that it's easier to process. Let's have a look at the data:

```
session.table("CAMPAIGN_INFO").show()
```

The preceding code shows the output of the CAMPAIGN_INFO table:

"ID"	"ACCEPTEDCMP1"	"ACCEPTEDCMP2"	"ACCEPTEDCMP3"	"ACCEPTEDCMP4"	"ACCEPTEDCMP5"	"RESPONSE"
5524	0	0	0	0	0	1
2174	0	0	0	0	0	0
4141	0	0	0	0	0	0
6182	0	0	0	0	0	0
5324	0	0	0	0	0	0
7446	0	0	0	0	0	0
965	0	0	0	0	0	0
6177	0	0	0	0	0	0
4855	0	0	0	0	0	1
5899	0	0	1	0	0	0

Figure 3.3 – The CAMPAIGN_INFO table

Loading and processing JSON files in Snowpark becomes easier when using the Variant column. Next, we will cover how to load a Parquet file into Snowflake using Snowpark.

Ingesting Parquet files into Snowflake

Parquet is a popular open source format for storing data licensed under Apache. The column-oriented format is lighter to store and faster to process. Parquet also supports complex data types since the data and the column information are stored in Parquet format. The COMPLAINT_INFO table consists of customer complaint information. Let's load this into Snowflake:

```
session.file.put('./datasets/complain_info.parquet', 'MY_STAGE')
```

The file will be uploaded into the internal stage. Snowpark can access it to process and load it into a table:

```
df_raw = session.read.parquet("@My_Stage/complain_info.parquet")
df_raw.copy_into_table("COMPLAINT_INFO")
```

The Parquet file is read into the DataFrame and then copied into the COMPLAINT_INFO table. Since the Parquet file already contains the table metadata information, it defines the table structure. We can query the table to view the data:

```
session.table("COMPLAINT_INFO").show()
```

This will output the following COMPLAINT_INFO table:

"ID"	"COMPLAIN"	"Z_COSTCONTACT"	"Z_REVENUE"
5524	0	3	11
2174	0	3	11
4141	0	3	11
6182	0	3	11
5324	0	3	11
7446	0	3	11
965	0	3	11
6177	0	3	11
4855	0	3	11
5899	0	3	11

Figure 3.4 – The COMPLAINT_INFO table

Parquet is one of the preferred formats for Snowflake since it's the format that's used by Apache Iceberg. Parquet stands out in data engineering and data science for its columnar storage, which optimizes compression and query performance. Its support for schema evolution and partitioning ensures flexibility and efficiency in handling evolving data structures. With broad compatibility across various data processing frameworks, Parquet enables seamless integration into existing workflows, making it a cornerstone format in modern data pipelines. In the next section, we will cover how easy it is to load unstructured data, such as an image, into Snowflake.

> **Important note**
>
> We've chosen to maintain separate stages for handling images and text, although it's not mandatory to do so. The MY_TEXT and MY_IMAGES stages can be prepared using the same methods we outlined earlier.

Ingesting images into Snowpark

Snowflake supports versatile data, such as images, that can be uploaded into a stage and executed directly in Snowpark without the need to manage dependencies as well.

Platforms such as Amazon S3, Google Cloud Storage, and Azure Blob Storage are commonly preferred for managing and storing image data due to their scalability and reliability. However, it's worth noting that Snowpark also offers flexibility in loading image files, making it a versatile option for handling image data in data engineering and data science workflows. We will be loading a bunch of sample images that can be used for processing:

```
session.file.put("./datasets/sample_images/*.png", "@My_Images")
```

The preceding code loads the images from the local folder to the internal stage. The path can support wildcard entries to upload all the images in a particular folder. The folder in the stage can be queried to get the list of images that were uploaded:

```
Session.sql("LS @My_Images").show()
```

The preceding code shows a list of all the images that are present in the stage:

```
---------------------------------------------------------------------------------------------------------
|"name"                 |"size" |"md5"                             |"last_modified"                |
---------------------------------------------------------------------------------------------------------
|my_images/100.png.gz   |2512   |9c413259f6c167cb8e6003a81fd49f57  |Wed, 19 Jul 2023 15:13:29 GMT  |
|my_images/1007.png.gz  |2288   |16b7222d78f72f87cd7b72aa41bc6a49  |Wed, 19 Jul 2023 15:13:29 GMT  |
|my_images/101.png.gz   |3040   |3fef4462ae2c75d4a7043f4f2bfa860a  |Wed, 19 Jul 2023 15:13:29 GMT  |
---------------------------------------------------------------------------------------------------------
```

Figure 3.5 – List of images

Once the image has been uploaded, it can be directly accessed via Snowpark. Snowpark supports the get_stream function to stream the file's contents as bytes from the stage. We can use a library such as Pillow to read the file from the bytes stream:

```
import PIL.Image

bytes_object = session.file.get_stream(
    "@My_Images/101.png.gz", decompress=True)
image = PIL.Image.open(bytes_object)
image.resize((150,150))
```

This will output the following image:

Figure 3.6 – Rendering images

The image is displayed directly in the notebook. Snowpark's native support for images supports capabilities for use cases such as image classification, image processing, and image recognition. Snowpark also supports rendering images dynamically. We will cover this in the next section.

Reading files dynamically with Snowpark

Snowpark contains the `files` module and the `SnowflakeFile` class, both of which provide access to files dynamically and stream them for processing. These dynamic files are also helpful for reading multiple files as we can iterate over them. `open()` extends the `IOBase` file objects and provides the functionality to open a file. The `SnowflakeFile` object also supports other `IOBase` methods for processing the file. The following code shows an example of reading multiple files using a relative path from the internal stage:

```
import snowflake.snowpark as snowpark
from snowflake.snowpark.functions import udf
from snowflake.snowpark.files import SnowflakeFile
from snowflake.snowpark.types import StringType, IntegerType

@udf(
    name="get_bytes_length",
    replace=True,
    input_types=[StringType()],
    return_type=IntegerType(),
    packages=['snowflake-snowpark-python']
)
def get_file_length(file_path):
    with SnowflakeFile.open(file_path) as f:
        s = f.read()
        return len(s)
```

The preceding code iterates over the `@MY_TEXTS` stage location and calculates the length of each file using the `SnowflakeFile` method. The path is passed as the input to the UDF. We can execute the function to get the output:

```
session.sql("SELECT RELATIVE_PATH, \
    get_bytes_length(build_scoped_file_url( \
        @MY_TEXTS,RELATIVE_PATH)) \
            as SIZE from DIRECTORY(@MY_TEXTS);").collect()
```

The preceding code produces the following result:

```
[Row(RELATIVE_PATH='text_2.txt', SIZE=738),
 Row(RELATIVE_PATH='text_3.txt', SIZE=761),
 Row(RELATIVE_PATH='text_1.txt', SIZE=1319)]
```

Figure 3.7 – Dynamic files within Snowpark

The files in the stage are displayed as output. In this section, we covered ingesting different types of files into Snowflake using Snowpark. In the next section, we will learn how to perform data preparation and transformations using Snowpark.

Data exploration and transformation

Once the data has been loaded, the next step is to prepare the data so that it can be transformed. In this section, we will cover how to perform data exploration so that we understand how the modify the data as necessary.

Data exploration

Data exploration is a critical step in data analysis as it sets the stage for successful insights and informed decision-making. By delving into the data, analysts can deeply understand its characteristics, uncover underlying patterns, and identify potential issues or outliers. Exploring the data provides valuable insights into its structure, distribution, and relationships, enabling analysts to choose the appropriate data transformation techniques.

Understanding the data's characteristics and patterns helps analysts determine the appropriate transformations and manipulations needed to clean, reshape, or derive new variables from the data. Additionally, data exploration aids in identifying subsets of data that are relevant to the analysis, facilitating the filtering and sub-setting operations required for specific analytical objectives.

Before embarking on data transformation, we must understand the data we have in place. By comprehensively understanding the data, we can effectively identify its structure, quality, and patterns. This understanding is a solid foundation for informed decision-making during the data transformation process, enabling us to extract meaningful insights and derive maximum value from the data. Take a look at the following code:

```
purchase_history = session.table("PURCHASE_HISTORY")
campaign_info = session.table("CAMPAIGN_INFO")
complain_info = session.table("COMPLAINT_INFO")
```

Here, we loaded the necessary tables into a session. These tables are now available in the Snowpark session for further data preparation. We will start by preparing the PURCHASE_HISTORY table:

```
purchase_history.show(n=5)
```

The `show()` method returns the data from the DataFrame. The preceding code produces the top 5 rows from the `PURCHASE_HISTORY` table:

```
--------------------------------------------------------------------------------------------------------------------------
|"ID"  |"YEAR_BIRTH"  |"EDUCATION"  |"MARITAL_STATUS"  |"INCOME"  |"KIDHOME"  |"TEENHOME"  |"DT_CUSTOMER"  |"RECENCY"
--------------------------------------------------------------------------------------------------------------------------
|5524  |1957          |Graduation   |Single            |58138     |0          |0           |2012-09-04     |58
|2174  |1954          |Graduation   |Single            |46344     |1          |1           |2014-03-08     |38
|4141  |1965          |Graduation   |Together          |71613     |0          |0           |2013-08-21     |26
|6182  |1984          |Graduation   |Together          |26646     |1          |0           |2014-02-10     |26
|5324  |1981          |PhD          |Married           |58293     |1          |0           |2014-01-19     |94
--------------------------------------------------------------------------------------------------------------------------
```

Figure 3.8 – PURCHASE_HISTORY – top 5 rows

We can use the `collect()` method to display the data in the notebook:

```
purchase_history.collect()
```

The records from the `PURCHASE_HISTORY` table are shown in the JSON array:

```
Row(ID=4047, YEAR_BIRTH=1954, EDUCATION='PhD', MARITAL_STATUS='Married', INCOME=65324, KIDHOME=0, TEENHOME=1, DT_CUSTOMER=datetime.date(2014, 1, 11), RECENCY
Row(ID=1409, YEAR_BIRTH=1951, EDUCATION='Graduation', MARITAL_STATUS='Together', INCOME=40689, KIDHOME=0, TEENHOME=1, DT_CUSTOMER=datetime.date(2013, 3, 18),
...
Row(ID=8492, YEAR_BIRTH=1978, EDUCATION='Graduation', MARITAL_STATUS='Together', INCOME=75437, KIDHOME=0, TEENHOME=0, DT_CUSTOMER=datetime.date(2013, 11, 9),
Row(ID=11188, YEAR_BIRTH=1957, EDUCATION='Graduation', MARITAL_STATUS='Together', INCOME=26091, KIDHOME=1, TEENHOME=1, DT_CUSTOMER=datetime.date(2014, 2, 25)
Row(ID=1212, YEAR_BIRTH=1973, EDUCATION='Graduation', MARITAL_STATUS='Married', INCOME=52845, KIDHOME=1, TEENHOME=0, DT_CUSTOMER=datetime.date(2013, 8, 13),
Row(ID=9097, YEAR_BIRTH=1956, EDUCATION='Graduation', MARITAL_STATUS='Divorced', INCOME=46086, KIDHOME=0, TEENHOME=1, DT_CUSTOMER=datetime.date(2013, 11, 3),
...]
```

Figure 3.9 – PURCHASE_HISTORY – full table

The difference between collect() and show()

In Snowpark Python, there are two essential functions: `collect()` and `show()`. These functions serve different purposes in processing and displaying data. The `collect()` function in Snowpark Python is used to gather or retrieve data from a specified source, such as a table, file, or API. It allows you to perform queries, apply filters, and extract the desired information from the data source. The collected data is stored in a variable or structure, such as a DataFrame, for further analysis or manipulation.

On the other hand, the `show()` function in Snowpark Python is primarily used to display the contents of a DataFrame or any other data structure in a tabular format. It provides a convenient way to visualize and inspect the data at different stages of the data processing pipeline. The `show()` function presents the data in a human-readable manner, showing the rows and columns in a structured table-like format. It can be helpful for debugging, understanding the data's structure, or performing exploratory data analysis.

In short, the `collect()` function focuses on gathering and retrieving data from a source, while the `show()` function displays the data in a readable format. Both functions play essential roles in Snowpark Python when it comes to working with data, but they serve distinct purposes in the data processing workflow.

Next, we will use the count () method to get the total count of the rows in the table:

```
purchase_history.count()
```

From the resulting output, we can see that the PURCHASE_HISTORY table contains around 2,000 rows of data.

We can now check the columns of the table to understand more about this data:

```
purchase_history.columns
```

This returns the column information, which helps us understand the data better. The column information contains the data related to customer purchase history:

```
['ID',
 'YEAR_BIRTH',
 'EDUCATION',
 'MARITAL_STATUS',
 'INCOME',
 'KIDHOME',
 'TEENHOME',
 'DT_CUSTOMER',
 'RECENCY',
 'MNTWINES',
 'MNTFRUITS',
 'MNTMEATPRODUCTS',
 'MNTFISHPRODUCTS',
 'MNTSWEETPRODUCTS',
 'MNTGOLDPRODS',
 'NUMDEALSPURCHASES',
 'NUMWEBPURCHASES',
 'NUMCATALOGPURCHASES',
 'NUMSTOREPURCHASES',
 'NUMWEBVISITSMONTH']
```

Figure 3.10 – PURCHASE_HISTORY columns

We can now filter the data to slice and dice it. We can use the following code to filter specific rows or a single row:

```
from snowflake.snowpark.functions import col
purchase_history.filter(col("id") == 1).show()
```

This returns the column. where id is set to 1. We can pass multiple values in the column filter to perform additional row-level operations:

"ID"	"YEAR_BIRTH"	"EDUCATION"	"MARITAL_STATUS"	"INCOME"	"KIDHOME"	"TEENHOME"	"DT_CUSTOMER"	"RECENCY"
1	1961	Graduation	Single	57091	0	0	2014-06-15	0

Figure 3.11 – PURCHASE_HISTORY ID filter

If we need to add multiple filter values, we can use the & operation to pass multiple column filter values to the method:

```
purchase_history.filter((col("MARITAL_STATUS") == "Married") &
                        (col("KIDHOME") == 1)).show()
```

The preceding code provides data about those with MARITAL_STATUS set to Married and who have kids at home (KIDHOME):

"ID"	"YEAR_BIRTH"	"EDUCATION"	"MARITAL_STATUS"	"INCOME"	"KIDHOME"	"TEENHOME"	"DT_CUSTOMER"	"RECENCY"
5324	1981	PhD	Married	58293	1	0	2014-01-19	94
6177	1985	PhD	Married	33454	1	0	2013-05-08	32
1994	1983	Graduation	Married	NULL	1	0	2013-11-15	11
9736	1980	Graduation	Married	41850	1	1	2012-12-24	51
5376	1979	Graduation	Married	2447	1	0	2013-01-06	42
2404	1976	Graduation	Married	53359	1	1	2013-05-27	4
9422	1989	Graduation	Married	38360	1	0	2013-05-31	26
10755	1976	2n Cycle	Married	23718	1	0	2013-09-02	76
503	1985	Master	Married	20559	1	0	2013-03-12	88
2139	1975	Master	Married	7500	1	0	2013-10-02	19

Figure 3.12 – PURCHASE_HISTORY filters

This helps us understand the purchase history pattern of married customers with kids. We can also filter it to the year of birth by passing the year of birth range between 1964 and 1980:

```
purchase_history.filter((col("YEAR_BIRTH") >= 1964) &
                        (col("YEAR_BIRTH") <= 1980)).show()
```

This displays the purchase data for customers born between 1964 and 1980:

"ID"	"YEAR_BIRTH"	"EDUCATION"	"MARITAL_STATUS"	"INCOME"	"KIDHOME"	"TEENHOME"	"DT_CUSTOMER"	"RECENCY"
4141	1965	Graduation	Together	71613	0	0	2013-08-21	26
7446	1967	Master	Together	62513	0	1	2013-09-09	16
965	1971	Graduation	Divorced	55635	0	1	2012-11-13	34
4855	1974	PhD	Together	30351	1	0	2013-06-06	19
387	1976	Basic	Married	7500	0	0	2012-11-13	59
9736	1980	Graduation	Married	41850	1	1	2012-12-24	51
5376	1979	Graduation	Married	2447	1	0	2013-01-06	42
7892	1969	Graduation	Single	18589	0	0	2013-01-02	89
2404	1976	Graduation	Married	53359	1	1	2013-05-27	4
1966	1965	PhD	Married	84618	0	0	2013-11-22	96

Figure 3.13 – PURCHASE_HISTORY filters

This data helps us understand their purchases. We can also use the select() method to select only the columns that are required for analysis:

```
purchase_history.select(col("ID"), col("YEAR_BIRTH"),
                        col("EDUCATION")).show()
```

The preceding returns only the customer's ID, year, and education status:

```
...  ------------------------------------
     |"ID"  |"YEAR_BIRTH"  |"EDUCATION"  |
     ------------------------------------
     |5524  |1957          |Graduation   |
     |2174  |1954          |Graduation   |
     |4141  |1965          |Graduation   |
     |6182  |1984          |Graduation   |
     |5324  |1981          |PhD          |
     |7446  |1967          |Master       |
     |965   |1971          |Graduation   |
     |6177  |1985          |PhD          |
     |4855  |1974          |PhD          |
     |5899  |1950          |PhD          |
     ------------------------------------
```

Figure 3.14 – PURCHASE_HISTORY columns

In the upcoming chapters, we will delve deeper into data exploration, uncovering more techniques to gain insights from our data.

Building upon these basic exploration steps, we will dive into the realm of data transformation operations. By combining our understanding of the data and the power of transformation techniques, we will unlock the full potential of our data and extract valuable insights for informed decision-making.

In the next section, we will discuss how to perform data transformation using this data.

Data transformations

Data transformation is a fundamental process that involves modifying and reshaping data to make it more suitable for analysis or other downstream tasks, such as the machine learning model building process. It entails applying a series of operations to the data, such as cleaning, filtering, aggregating, and reformatting, to ensure its quality, consistency, and usability. Data transformation allows us to convert raw data into a structured and organized format that can be easily interpreted and analyzed.

The data requires minimal transformation, and we will cover it extensively in the coming chapters. Our goal for this section is to combine data from different sources, creating a unified table for further processing that we will use in the next chapter. We will leverage Snowpark's robust join and union capabilities to accomplish this. By utilizing joins, we can merge data based on standard columns or conditions. Unions, on the other hand, allow us to append data from multiple sources vertically. These techniques will enable us to integrate and consolidate our data efficiently, setting the stage for comprehensive analysis and insights. Let's explore how Snowpark's join and union capabilities can help us achieve this data combination:

```
purchase_campaign = purchase_history.join(
    campaign_info,
```

```
        purchase_history.ID == campaign_info.ID ,
        lsuffix="_left", rsuffix="_right"
    )
```

Here, we are joining the purchase history to campaign information to establish the relationship between purchases and campaigns. The standard ID column is used to select the join and defaults to an inner join:

```
purchase_campaign = purchase_campaign.drop("ID_RIGHT")
```

We are dropping the extra ID column from the joined result. The DataFrame now contains just a single ID column:

```
purchase_campaign.show()
```

This displays the data of the purchase campaign combined with the purchase history and the campaign information:

"ID_LEFT"	"YEAR_BIRTH"	"EDUCATION"	"MARITAL_STATUS"	"INCOME"	"KIDHOME"	"TEENHOME"	"DT_CUSTOMER"	"RECENCY"
5524	1957	Graduation	Single	58138	0	0	2012-09-04	58
2174	1954	Graduation	Single	46344	1	1	2014-03-08	38
4141	1965	Graduation	Together	71613	0	0	2013-08-21	26
6182	1984	Graduation	Together	26646	1	0	2014-02-10	26
5324	1981	PhD	Married	58293	1	0	2014-01-19	94
7446	1967	Master	Together	62513	0	1	2013-09-09	16
965	1971	Graduation	Divorced	55635	0	1	2012-11-13	34
6177	1985	PhD	Married	33454	1	0	2013-05-08	32
4855	1974	PhD	Together	30351	1	0	2013-06-06	19
5899	1950	PhD	Together	5648	1	1	2014-03-13	68

Figure 3.15 – Purchase campaign data

Let's combine this with the complaint information to get the complete data:

```
final_combined = purchase_campaign.join(
    complain_info,
    purchase_campaign["ID_LEFT"] == complain_info.ID
)
final_combined = final_combined.drop("ID_LEFT")
```

Here, we are combining the result of the purchase campaign along with the complaint information by using the standard ID column. The resultant DataFrame contains the complete data required for data analysis. We are dropping the extra ID column from the joined result. The DataFrame now has just a single ID column:

```
final_combined.show()
```

This displays the final data combined from all three tables. We can now write this data into the table for further analysis:

```
final_combined.write.save_as_table("MARKETING_DATA")
```

Here, the data is written into the MARKETING_DATA table, at which point it will be available inside Snowflake. We need to append this data with the additional marketing data that must be loaded into this table.

The difference between joins and unions

Joins combine data from two or more tables based on a shared column or condition. In Snowflake Snowpark, you can perform different types of joins, such as inner join, left join, right join, and full outer join. Joins allow you to merge data horizontally by aligning rows based on matching values in the specified columns. This enables you to combine related data from multiple tables, resulting in a combined dataset that includes information from all the joined tables.

On the other hand, unions are used to append data from multiple tables vertically, or result sets into a single dataset. Unlike joins, unions do not require any specific conditions or matching columns. Instead, they stack rows on top of each other, concatenating the data vertically. This is useful when you have similar datasets with the same structure and want to consolidate them into a single dataset. Unions can be performed in Snowflake Snowpark to create a new dataset that contains all the rows from the input tables or result sets.

In summary, joins in Snowflake Snowpark are used to combine data horizontally by matching columns, while unions are used to stack data vertically without any specific conditions. Joins merge related data from multiple tables, while unions append similar datasets into a single dataset.

Appending data

The Snowflake Snowpark UNION function is vital in combining and integrating new data into a Snowflake database. The importance of the UNION function lies in its ability to append rows from different data sources vertically, or result sets into a single consolidated dataset. When new data is added to the database, it is often necessary to merge or combine it with existing data for comprehensive analysis. The UNION function allows us to seamlessly integrate the newly added data with the existing dataset, creating a unified view encompassing all relevant information.

This capability of the UNION function is precious in scenarios where data is received or updated periodically. For example, suppose we receive daily sales data or log files. In that case, the UNION function enables us to effortlessly append the new records to the existing dataset, ensuring that our analysis reflects the most up-to-date information. Additionally, it ensures data consistency and allows for seamless continuity in data analysis, enabling us to derive accurate insights and make informed decisions based on the complete and unified dataset.

The additional marketing data is available in the MARKETING_ADDITIONAL table. Let's see how we can leverage Snowpark's UNION function to include this additional data for processing:

```
marketing_additional = session.table("MARKETING_ADDITIONAL")
marketing_additional.show()
```

The preceding code displays the data from the MARKETING_ADDITIONAL table:

"ID"	"YEAR_BIRTH"	"EDUCATION"	"MARITAL_STATUS"	"INCOME"	"KIDHOME"	"TEENHOME"	"DT_CUSTOMER"	"RECENCY"
4860	1970	Graduation	Single	24206	1	0	2013-03-08	66
10757	1967	PhD	Divorced	28420	1	0	2013-12-24	36
4023	1970	Graduation	Married	22979	1	0	2012-09-06	29
6679	1966	Graduation	Single	33279	0	0	2014-06-12	29
9923	1972	Master	Single	46423	1	1	2013-09-18	6
7181	1977	Graduation	Married	30368	0	1	2013-11-07	97

Figure 3.16 – The MARKETING_ADDITIONAL table

With that, the table has been loaded into the DataFrame. Let's look at the number of rows in our original and appended tables:

```
print("No of rows in MARKETING_ADDITIONAL table: \
    ",marketing_additional.count())
print("No of rows in PURCHASE_HISTORY table: \
    ",final_combined.count())
```

This code displays the total number of rows in the MARKETING_ADDITIONAL and PURCHASE_HISTORY tables:

```
No of rows in MARKETING_ADDITIONAL table:  240
No of rows in PURCHASE_HISTOTY table:  2000
```

Figure 3.17 – Data row count

The MARKETING_ ADDITIONAL table contains 240 rows of new data that must be appended with the PURCHASE_HISTORY table, which contains 2,000 rows of data. Since the column names are identical, the data can be appended by using union_by_name:

```
final_appended = final_combined.union_by_name(marketing_additional)
```

Now, the DataFrame contains the appended data. Let's look at the number of rows in this DataFrame:

```
print("No of rows in UPDATED table: ",final_appended.count())
final_appended.show()
```

The preceding code shows the final data that's in the DataFrame:

"YEAR_BIRTH"	"EDUCATION"	"MARITAL_STATUS"	"INCOME"	"KIDHOME"	"TEENHOME"	"DT_CUSTOMER"	"RECENCY"	"MNTWINES"
1957	Graduation	Single	58138	0	0	2012-09-04	58	635
1954	Graduation	Single	46344	1	1	2014-03-08	38	11
1965	Graduation	Together	71613	0	0	2013-08-21	26	426
1984	Graduation	Together	26646	1	0	2014-02-10	26	11
1981	PhD	Married	58293	1	0	2014-01-19	94	173
1967	Master	Together	62513	0	1	2013-09-09	16	520
1971	Graduation	Divorced	55635	0	1	2012-11-13	34	235
1985	PhD	Married	33454	1	0	2013-05-08	32	76

No of rows in UPDATED table: 2240

Figure 3.18 – The MARKETING_FINAL table

The total count of the rows is 2,240. With that, the new data has been appended. Now, we will write this data into the MARKETING_FINAL table in Snowflake:

```
final_appended.write.save_as_table("MARKETING_FINAL")
```

The MARKETING_DATA table is now available in Snowflake and can be consumed.

> **The difference between union and union_by_name**
>
> Two methods are available for combining data: union_by_name and union. Both methods allow multiple datasets to be merged, but they differ in their approach and functionality.
>
> The union_by_name method in Snowpark Python is specifically designed to combine datasets by matching and merging columns based on their names. This method ensures that the columns with the same name from different datasets are merged, creating a unified dataset. It is beneficial when you have datasets with similar column structures and want to consolidate them while preserving the column names.
>
> On the other hand, the union method in Snowpark Python combines datasets by simply appending them vertically, regardless of column names or structures. This method concatenates the rows from one dataset with the rows from another, resulting in a single dataset with all the rows from both sources. The union method is suitable for stacking datasets vertically without considering column names or matching structures. However, note that in certain cases, the column type matters, such as when casting a string column to a numeric value.

Data grouping and analysis

Now that the data is ready and has been transformed, the next step is to see how we can group data to understand important patterns and analyze it. In this section, we will aggregate this data and analyze it.

Data grouping

In data analysis, understanding patterns within datasets is crucial for gaining insights and making informed decisions. One powerful tool that aids in this process is the group_by function in Snowpark Python. This function allows us to group data based on specific criteria, enabling us to dissect and analyze the dataset in a structured manner.

By utilizing the group_by function, we can uncover valuable insights into how data is distributed and correlated across different categories or attributes. For example, we can group sales data by product category to analyze sales trends, or group customer data by demographics to understand buying behavior.

Furthermore, the group_by function can be combined with other data manipulation and visualization techniques to gain deeper insights. For instance, we can create visualizations such as bar charts or heatmaps to visually represent the aggregated data, making it easier to spot patterns and trends.

To facilitate grouping and conducting deeper analysis, we'll utilize the MARKETING_FINAL table we established earlier:

```
marketing_final = session.table("MARKETING_FINAL")
```

Here, we are loading the data from the MARKETING_FINAL table into the DataFrame. We will use this DataFrame to perform aggregations:

```
marketing_final.group_by("EDUCATION").mean("INCOME").show()
```

This returns the average income by EDUCATION. People with PhDs have the highest average income, and people with primary education have the lowest average income:

Figure 3.19 – Average income by education

Now, we can create an alias for the column:

```
marketing_final.group_by("EDUCATION").agg(avg("INCOME").alias( \
    "Avg_Income")).show()
```

The average income is displayed as an alias – `AVG_INCOME`:

"EDUCATION"	"AVG_INCOME"
Graduation	52720.373656
PhD	56145.313929
Master	52917.534247
Basic	20306.259259
2n Cycle	47633.190000

Figure 3.20 – The AVG_INCOME alias

We can also achieve similar results by using the `function()` method to pass the respective operation from Snowpark functions:

```
marketing_final.group_by("MARITAL_STATUS").function("sum")( \
    "Z_REVENUE").show()
```

This prints the following output:

"MARITAL_STATUS"	"SUM(Z_REVENUE)"
Single	5280
Together	6380
Married	9504
Divorced	2552
Widow	847
Alone	33
Absurd	22
YOLO	22

Figure 3.21 – Sum of revenue by marital status

Here, we can see that married customers generate the highest revenue. We can also use `agg()` to perform this particular aggregation. Let's calculate the maximum income by marital status:

```
marketing_final.group_by("MARITAL_STATUS").agg(max("INCOME")).show()
```

This generates the following output:

```
--------------------------------------
|"MARITAL_STATUS"   |"MAX(INCOME)"   |
--------------------------------------
|Single             |113734          |
|Together           |666666          |
|Married            |160803          |
|Divorced           |153924          |
|Widow              |85620           |
|Alone              |61331           |
|Absurd             |79244           |
|YOLO               |48432           |
--------------------------------------
```

Figure 3.22 – Income by marital status

Here, we can see that customers who are together and married as a family have the maximum income to spend, and hence they generate the maximum revenue. Next, we will find the count of different types of graduates and their maximum income:

```
marketing_final.group_by("EDUCATION").agg((col("*"), "count"),
    max("INCOME")).show()
```

The preceding code produces the following output:

```
-------------------------------------------------------
|"EDUCATION"  |"COUNT(LITERAL())"  |"MAX(INCOME)"   |
-------------------------------------------------------
|Graduation   |1127                |666666          |
|PhD          |486                 |162397          |
|Master       |370                 |157733          |
|Basic        |54                  |34445           |
|2n Cycle     |203                 |96547           |
-------------------------------------------------------
```

Figure 3.23 – Count of category

Here, we can see that `PhD` has a maximum income of `162397`, and that people with `Basic` income have the lowest maximum income – that is, `34445`.

We can also perform complex multi-level aggregations in Snowpark. Let's find out how people with different educations and marital statuses spend:

```
marketing_final.group_by(["EDUCATION","MARITAL_STATUS"]).agg(
    avg("INCOME").alias("Avg_Income"),
    sum("NUMSTOREPURCHASES").alias("Sum_Purchase")
).show()
```

```
|"EDUCATION"   |"MARITAL_STATUS"   |"AVG_INCOME"    |"SUM_PURCHASE"  |
-------------------------------------------------------------------------
|Graduation    |Single             |51322.182927    |1426            |
|Graduation    |Together           |55758.480702    |1686            |
|PhD           |Married            |58138.031579    |1191            |
|Master        |Together           |52109.009804    |600             |
|Graduation    |Divorced           |54526.042017    |697             |
|Graduation    |Married            |50800.258741    |2544            |
|2n Cycle      |Single             |53673.944444    |194             |
|Master        |Married            |53286.028986    |815             |
|Graduation    |Widow              |54976.657143    |206             |
|2n Cycle      |Married            |46201.100000    |440             |
-------------------------------------------------------------------------
```

Figure 3.24 – Multi-level aggregation

Let's determine the relationship between EDUCATION, MARITAL_STATUS, and SUM_PURCHASE. People who are graduates and married spend the most compared to single people. We can also sort the results by using the sort() function:

```
aggregate_result = marketing_final.group_by(["EDUCATION","MARITAL_
STATUS"]).agg(
    avg("INCOME").alias("Avg_Income"),
    sum("NUMSTOREPURCHASES").alias("Sum_Purchase")
)
aggregate_result.sort(
    col("EDUCATION").asc(), col("Sum_Purchase").asc()
).show()
```

Here's the output:

```
|"EDUCATION"   |"MARITAL_STATUS"   |"AVG_INCOME"    |"SUM_PURCHASE"  |
-------------------------------------------------------------------------
|2n Cycle      |Widow              |51392.200000    |37              |
|2n Cycle      |Divorced           |49395.130435    |138             |
|2n Cycle      |Single             |53673.944444    |194             |
|2n Cycle      |Together           |44736.410714    |309             |
|2n Cycle      |Married            |46201.100000    |440             |
|Basic         |Widow              |22123.000000    |3               |
|Basic         |Divorced           |9548.000000     |3               |
|Basic         |Together           |21240.071429    |34              |
|Basic         |Single             |18238.666667    |49              |
|Basic         |Married            |21960.500000    |65              |
-------------------------------------------------------------------------
```

Figure 3.25 – Sorted result

Here, we are sorting the results in ascending order by purchase amount after the aggregation is completed. The following section will cover some standard data analysis that can be performed on this data.

Data analysis

In the previous sections, we delved into data exploration, transformation, and aggregation, where we learned about various techniques we can use to find out what our data is all about and how we can combine different datasets. Armed with a solid foundation of general dataset exploration, we are ready to dive deeper into data analysis using Snowpark Python.

This section focuses on leveraging the power of statistical functions, sampling techniques, pivoting operations, and converting data into a pandas DataFrame for advanced analysis. We will explore applying statistical functions to extract meaningful information from our data. Then, we will learn about different sampling techniques to work efficiently with large datasets. Additionally, we will discover how to reshape our data using pivoting operations to facilitate in-depth analysis.

Moreover, we will explore the seamless integration of Snowpark Python with pandas, a widely used data manipulation library. We will understand how to convert our Snowpark data into a pandas DataFrame, enabling us to leverage pandas' extensive analytical and visualization capabilities.

The following section provides a glimpse into the capabilities of Snowpark Python for data analysis; we will delve deeper into each topic in the subsequent chapter. Here, we aim to provide a foundational understanding of the key concepts and techniques of analyzing data using Snowpark Python. In the next chapter, we will explore these topics in greater detail, unraveling the full potential of Snowpark Python for data analysis.

Describing the data

The first step in our analysis is understanding how our data is distributed. The describe() function in pandas is a valuable tool that helps us gain insights into the statistical properties of our numerical data. When we apply describe() to a DataFrame, it computes various descriptive statistics, including the count, mean, standard deviation, minimum, quartiles, and maximum values for each numerical column.

This summary comprehensively overviews our data's distribution and central tendencies. By examining these statistics, we can quickly identify key characteristics, such as the range of values, the spread of the data, and any potential outliers. This initial exploration sets the stage for more advanced analysis techniques and allows us to make informed decisions based on a solid understanding of our dataset's distribution:

```
marketing_final.describe().show()
```

The preceding code shows the data from the MARKETING_FINAL table:

"SUMMARY"	"YEAR_BIRTH"	"EDUCATION"	"MARITAL_STATUS"	"INCOME"	"KIDHOME"	"TEENHOME"
count	2240.0	2240	2240	2216.0	2240.0	2240.0
mean	1968.805804	NULL	NULL	52247.251354	0.444196	0.50625
stddev	11.984069467422158	NULL	NULL	25173.076660901403	0.5383985512610523	0.544538336575121
min	1893.0	2n Cycle	Absurd	1730.0	0.0	0.0
max	1996.0	PhD	YOLO	666666.0	2.0	2.0

Figure 3.26 – MARKETING_FINAL DataFrame

The result shows the different columns and the data in the MARKETING_FINAL table.

Finding distinct data

In Snowpark DataFrames, the distinct() function is crucial in identifying unique values within a column or set of columns. When applied to a Snowpark DataFrame, distinct() eliminates duplicate records, resulting in a new DataFrame that contains only distinct values. This function is particularly useful for dealing with large datasets or extracting unique records for analysis or data processing:

```
marketing_final.distinct().count()
```

The preceding code shows the total count of the MARKETING_FINAL table:

Figure 3.27 – MARKETING_FINAL count

In our case, the entire dataset is returned since we do not have any duplicate rows. distinct() preserves the original rows of the DataFrame and only filters out repeated values within the specified columns.

Dropping duplicates

drop_duplicates() removes duplicate rows from a Snowpark DataFrame. It analyzes the entire row and compares it with other rows in the DataFrame. If a row is found to be an exact duplicate of another row, drop_duplicates() will remove it, keeping only the first occurrence. By default, this function considers all columns in the DataFrame for duplicate detection:

```
marketing_final.select(["Education","Marital_Status"]).drop_
duplicates().show()
```

This will display the following output:

```
...    -----------------------------------+-+
       |"EDUCATION"    |"MARITAL_STATUS"   | |
       -----------------------------------+-+
       |Graduation     |Single             | |
       |Graduation     |Together           | |
       |PhD            |Married            | |
       |Master         |Together           | |
       |Graduation     |Divorced           | |
       |Graduation     |Married            | |
       |Basic          |Married            | |
       |2n Cycle       |Single             | |
       |Master         |Married            | |
       |Master         |Single             | |
       -----------------------------------+-+
```

Figure 3.28 – Marketing duplicates removed

Note that you can specify specific columns using the `subset` parameter to check for duplicates based on those columns alone. `drop_duplicates()` modifies the original DataFrame by removing duplicate rows.

Crosstab analysis

Once we have identified the unique combinations of the EDUCATION and MARITAL_STATUS columns in our dataset, we might still be curious about how frequently each combination occurs. We can utilize the `crosstab` function to determine the occurrence of these unique combinations. By applying the `crosstab` function to our dataset, we can generate a cross-tabulation or contingency table that displays the frequency distribution of the unique combinations of EDUCATION and MARITAL_STATUS:

```
marketing_final.stat.crosstab(col1="Education",col2="Marital_Status").
show()
```

The preceding code shows the crosstab data in the DataFrame:

"EDUCATION"	"'Single'"	"'Together'"	"'Married'"	"'Divorced'"	"'Widow'"	"'Alone'"	"'Absurd'"	"'YOLO'"
Graduation	252	286	433	119	35	1	1	0
PhD	98	117	192	52	24	1	0	2
Master	75	106	138	37	12	1	1	0
Basic	18	14	20	1	1	0	0	0
2n Cycle	37	57	81	23	5	0	0	0

Figure 3.29 – Crosstab data

This table provides a comprehensive overview of how often each unique combination occurs in the dataset, allowing us to gain valuable insights into the relationships between these variables. The `crosstab` function aids us in understanding the distribution and occurrence patterns of the unique combinations, further enhancing our data analysis capabilities.

Pivot analysis

Upon using the `crosstab` function to examine the unique combinations of the EDUCATION and MARITAL_STATUS columns in our dataset, we might encounter certain combinations with zero occurrences. We can construct a pivot table to gain a more comprehensive understanding of the data and further investigate the relationships between these variables.

Constructing a pivot table allows us to summarize and analyze the data more dynamically and flexibly. Unlike the `crosstab` function, which only provides the frequency distribution of unique combinations, a pivot table allows us to explore additional aggregate functions, such as sum, average, or maximum values. This enables us to delve deeper into the dataset and obtain meaningful insights:

```
market_subset = marketing_final.select(
    "EDUCATION","MARITAL_STATUS","INCOME"
)
market_pivot = market_subset.pivot(
    "EDUCATION",
    ["Graduation","PhD","Master","Basic","2n Cycle"]
).sum("INCOME")
market_pivot.show()
```

The preceding code shows the data in the DataFrame:

"MARITAL_STATUS"	"'Graduation'"	"'PhD'"	"'Master'"	"'Basic'"	"'2n Cycle'"
Single	12625257	5118203	4014792	328296	1932262
Together	15891167	6500805	5315119	297361	2505239
Married	21793311	11046226	7353472	439210	3696088
Divorced	6488599	2761024	1862282	9548	1136088
Widow	1924183	1446914	642417	22123	256961
Alone	34176	35860	61331	NULL	NULL
Absurd	79244	NULL	65487	NULL	NULL
YOLO	NULL	96864	NULL	NULL	NULL

Figure 3.30 – Pivot table

By constructing a pivot table for the EDUCATION and MARITAL_STATUS columns, we can uncover the occurrence counts and various statistical measures or calculations associated with each combination. This expanded analysis provides a more comprehensive view of the data and allows for a more nuanced and detailed exploration.

> **Note**
>
> When the `crosstab` function displays zero occurrences for certain combinations of variables, it is essential to note that those combinations will be represented as NULL values instead of zeros when constructing a pivot table.
>
> Unlike `crosstab`, which explicitly highlights zero counts for combinations absent in the dataset, a pivot table considers all possible combinations of the variables. Consequently, if a variety does not exist in the dataset, the corresponding cell in the pivot table will be represented as a NULL value rather than a zero.
>
> The presence of NULL values in the pivot table highlights the absence of data for those particular combinations. Interpreting and handling these NULL values appropriately during subsequent data analysis processes, such as data cleaning, imputation, or further statistical calculations, is essential.

Dropping missing values

The `dropna()` function in pandas is a powerful tool for handling missing values in a DataFrame. In this case, we will be utilizing the `dropna()` functionality of Snowpark, which allows us to remove rows or columns that contain missing or NULL values, helping to ensure the integrity and accuracy of our data. The `dropna()` function offers several parameters that provide flexibility in controlling the operation's behavior:

```
market_pivot.dropna(how="all").show()
```

The preceding code shows the data with the applied filter from the DataFrame:

"MARITAL_STATUS"	"'Graduation'"	"'PhD'"	"'Master'"	"'Basic'"	"'2n Cycle'"
Single	12625257	5118203	4014792	328296	1932262
Together	15891167	6500805	5315119	297361	2505239
Married	21793311	11046226	7353472	439210	3696088
Divorced	6488599	2761024	1862282	9548	1136088
Widow	1924183	1446914	642417	22123	256961
Alone	34176	35860	61331	NULL	NULL
Absurd	79244	NULL	65487	NULL	NULL
YOLO	NULL	96864	NULL	NULL	NULL

Figure 3.31 – Pivot table – dropna()

The how parameter determines the criteria that are used to drop rows or columns. It accepts the input as any and all: any drops the row or column if it contains any missing value, and all drops the row or column only if all its values are missing.

The `thresh` parameter specifies the minimum number of non-null values required to keep a row or column. The row or column is dropped if the *non-null values exceed* the threshold:

```
market_pivot.dropna(thresh=5).show()
```

The preceding code shows the data with the applied filter from the DataFrame:

"MARITAL_STATUS"	"'Graduation'"	"'PhD'"	"'Master'"	"'Basic'"	"'2n Cycle'"
Single	12625257	5118203	4014792	328296	1932262
Together	15891167	6500805	5315119	297361	2505239
Married	21793311	11046226	7353472	439210	3696088
Divorced	6488599	2761024	1862282	9548	1136088
Widow	1924183	1446914	642417	22123	256961

Figure 3.32 – Pivot threshold

The `subset` parameter allows us to specify a subset of columns or rows for missing value removal. It accepts a list of column or row labels. By default, `dropna()` checks all columns or rows for missing values. However, with a subset, we can focus on specific columns or rows for the operation:

```
market_pivot.dropna(subset="'Graduation'").show()
```

The preceding code drops any rows from the `market_pivot` DataFrame where the `Graduation` column has missing values and then displays the resulting DataFrame:

"MARITAL_STATUS"	"'Graduation'"	"'PhD'"	"'Master'"	"'Basic'"	"'2n Cycle'"
Single	12625257	5118203	4014792	328296	1932262
Together	15891167	6500805	5315119	297361	2505239
Married	21793311	11046226	7353472	439210	3696088
Divorced	6488599	2761024	1862282	9548	1136088
Widow	1924183	1446914	642417	22123	256961
Alone	34176	35860	61331	NULL	NULL
Absurd	79244	NULL	65487	NULL	NULL

Figure 3.33 – Pivot subset

This shows the data with the applied filter from the DataFrame.

> **Note**
>
> When working with pivot tables, it is crucial to handle NULL values appropriately because they can impact the accuracy and reliability of subsequent analyses. This allows us to ensure that we have complete data for further analysis and calculations.
>
> Having NULL values in the pivot result can lead to incorrect interpretations or calculations since NULL values can propagate through the analysis and affect subsequent aggregations, statistics, or visualizations. By replacing NULL values with a specific value, such as 0, we can provide a meaningful representation of the data in the pivot table, allowing us to perform reliable analysis and make informed decisions based on complete information.

Filling missing values

The fillna() function allows us to replace null values with specific values or apply various techniques for imputation. It also allows us to fill in the missing values in a DataFrame, ensuring that we maintain the integrity of the data structure. We can specify the values for filling nulls, such as a constant value, or values derived from statistical calculations such as mean, median, or mode. The fillna() function is useful when we're treating null values while considering the data's nature and the desired analysis:

```
market_pivot.fillna(0).show()
```

The preceding code fills any null values in the market_pivot DataFrame with a value of 0 and then displays the resulting DataFrame:

```
----------------------------------------------------------------------------------------
|"MARITAL_STATUS"  |"'Graduation'"  |"'PhD'"   |"'Master'"  |"'Basic'"  |"'2n Cycle'"  |
----------------------------------------------------------------------------------------
|Single            |12625257        |5118203   |4014792     |328296     |1932262       |
|Together          |15891167        |6500805   |5315119     |297361     |2505239       |
|Married           |21793311        |11046226  |7353472     |439210     |3696088       |
|Divorced          |6488599         |2761024   |1862282     |9548       |1136088       |
|Widow             |1924183         |1446914   |642417      |22123      |256961        |
|Alone             |34176           |35860     |61331       |0          |0             |
|Absurd            |79244           |0         |65487       |0          |0             |
|YOLO              |0               |96864     |0           |0          |0             |
----------------------------------------------------------------------------------------
```

Figure 3.34 – Missing values

This is a handy function that fills in missing values that need to be used for calculations.

Variable interaction

The `corr()` function calculates the correlation coefficient, which measures the strength and direction of the linear relationship between two variables. It returns a value between -1 and 1, where -1 represents a perfect negative correlation, 1 illustrates a perfect positive correlation, and 0 indicates no linear correlation:

```
marketing_final.stat.corr("INCOME", "NUMSTOREPURCHASES")
```

By executing this code, we obtain the correlation coefficient between the INCOME and NUMSTOREPURCHASES columns, providing insights into the potential relationship between income levels and the number of store purchases in the dataset:

```
0.5293621402734197
```

Figure 3.35 – Correlation value

The `cov()` function, on the other hand, calculates the covariance, which measures the degree of association between two variables without normalizing for scale:

```
marketing_final.stat.cov("INCOME", "NUMSTOREPURCHASES")
```

Here's the output:

```
43318.897447865325
```

Figure 3.36 – Covariance value

The covariance between the INCOME and NUMSTOREPURCHASES columns helps us understand how changes in income levels correspond to changes in the number of store purchases in the dataset.

> **Note**
>
> While both `corr()` and `cov()` help analyze relationships between variables, it is essential to note that in Snowpark Python, these functions only support the analysis of two variables at a time. This limitation means we can only calculate the correlation or covariance between two columns in a DataFrame, and not simultaneously across multiple variables. Additional techniques or functions may be required to overcome this limitation and perform correlation or covariance analysis for various variables.

Operating with pandas DataFrame

Converting a Snowpark DataFrame into a pandas DataFrame is a valuable step that opens up a wide range of analysis capabilities. Snowpark provides seamless integration with pandas, allowing us to leverage pandas' extensive data manipulation, analysis, and visualization functionalities. By converting a Snowpark DataFrame into a pandas DataFrame, we gain access to a vast ecosystem of tools and libraries that are designed explicitly for data analysis.

This transition enables us to leverage pandas' rich functions and methods, such as statistical calculations, advanced filtering, grouping operations, and time series analysis. pandas also provide many visualization options, such as generating insightful plots, charts, and graphs that are more accessible, to visualize the data. With pandas, we can create meaningful visual representations of our data, facilitating the exploration of patterns, trends, and relationships. Additionally, working with pandas allows us to utilize its extensive community support and resources. The pandas library has a vast user community, making finding documentation, tutorials, and helpful discussions on specific data analysis tasks more accessible.

Limitations of pandas DataFrames

Converting a Snowpark DataFrame into a pandas DataFrame can have its limitations, mainly when dealing with large datasets. The primary constraint is memory consumption as converting the entire dataset simultaneously may exceed available memory resources. This can hinder the analysis process and potentially lead to system crashes or performance issues.

However, these limitations can be mitigated by breaking the DataFrame into batches and sampling the data. We'll discuss this shortly.

Data analysis using pandas

Converting a Snowpark DataFrame into a pandas DataFrame empowers us to seamlessly transition from Snowpark's powerful data processing capabilities to pandas' feature-rich environment. This interoperability expands our analytical possibilities and enables us to perform advanced analysis and gain deeper insights from our data:

```
pandas_df = marketing_final.to_pandas()
pandas_df.head()
```

The preceding code converts the `marketing_final` Snowpark DataFrame into a pandas DataFrame, allowing us to work with the data using pandas' extensive data analysis and manipulation functionalities. It will print out the following output:

	YEAR BIRTH	EDUCATION	MARITAL STATUS	INCOME	KIDHOME	TEENHOME	DT CUSTOMER	RECENCY	MNTWINES
0	1957	Graduation	Single	58138.0	0	0	2012-09-04	58	635
1	1954	Graduation	Single	46344.0	1	1	2014-03-08	38	11
2	1965	Graduation	Together	71613.0	0	0	2013-08-21	26	426
3	1984	Graduation	Together	26646.0	1	0	2014-02-10	26	11
4	1981	PhD	Married	58293.0	1	0	2014-01-19	94	173

5 rows × 29 columns

Figure 3.37 – The resulting pandas DataFrame

This shows the data that has been converted into the pandas DataFrame.

Correlation in pandas

In pandas, calculating correlations among multiple columns is straightforward: it involves selecting the desired columns and applying the `corr()` function. It generates a correlation matrix, allowing us to examine the relationships between each pair of columns simultaneously:

```
pandas_df[["INCOME","KIDHOME","RECENCY"]].corr()
```

The preceding code calculates the correlation matrix among the INCOME, KIDHOME, and RECENCY columns in the pandas_df pandas DataFrame. It computes the pairwise correlation coefficients between these columns, providing insights into their relationships. The output is as follows:

	INCOME	KIDHOME	RECENCY
INCOME	1.000000	-0.428669	-0.003970
KIDHOME	-0.428669	1.000000	0.008827
RECENCY	-0.003970	0.008827	1.000000

Figure 3.38 – Pandas correlation

Next, we'll look at frequency distribution.

Frequency distribution

Calculating the frequency of values in a single column is simpler in pandas than in Snowpark Python. We can quickly obtain the frequency distribution in pandas by using the `value_counts()` function on a specific column. It returns a Series with unique values as indices and their corresponding counts as values. This concise method allows us to quickly understand the distribution and prevalence of each unique value in the column. On the other hand, in Snowpark Python, obtaining the frequency of values in a single column requires more steps and additional coding. We typically need to group the DataFrame by the desired column and then perform aggregation operations to count the occurrences of each unique value. Although this can be achieved in Snowpark Python, it involves more complex syntax and multiple transformations, making the process more cumbersome compared to pandas:

```
frequency = pandas_df.EDUCATION.value_counts()
frequency
```

`frequency = pandas_df.EDUCATION.value_counts()` calculates the frequency distribution of unique values in the `EDUCATION` column of the `pandas_df` pandas DataFrame and assigns the result to the `frequency` variable. The output is as follows:

```
Graduation      1127
PhD              486
Master           370
2n Cycle         203
Basic             54
Name: EDUCATION, dtype: int64
```

Figure 3.39 – Pandas data frequency

This shows the data frequency values in the pandas DataFrame.

Visualization in pandas

Creating visualizations is made easy with pandas due to its seamless integration with popular visualization libraries such as Matplotlib and Seaborn. pandas provides a simple and intuitive interface to generate various visualizations, including line plots, bar charts, histograms, scatter plots, and more.

By leveraging pandas' built-in plotting functions, we can effortlessly transform our data into insightful visual representations, enabling us to explore patterns, trends, and relationships within our dataset. With just a few lines of code, pandas *empowers* us to produce visually appealing and informative plots, facilitating the communication and interpretation of our data:

```
frequency.plot(kind="barh",figsize=(8,3))
```

The preceding code creates a horizontal bar plot from the frequency distribution data stored in the `frequency` variable, where each unique value is represented by a bar with a length proportional to its count, and the plot has a customized size of 8 inches in width and 3 inches in height:

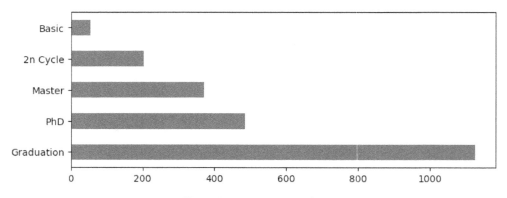

Figure 3.40 – Frequency plot

Similarly, we can generate a Hexbin plot by changing `kind` to `hexbin`:

```
pandas_df.plot(
    kind="hexbin",
    x="INCOME",y="MNTGOLDPRODS",
    xlim=[0,100000],ylim=[0,100],
    figsize=(8,3)
)
```

The preceding code creates a Hexbin plot that visualizes the relationship between the INCOME and MNTGOLDPRODS columns in the `pandas_df` pandas DataFrame:

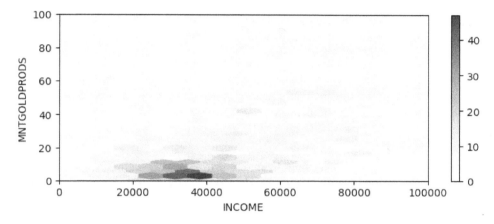

Figure 3.41 – Hexbin plot

Here, the *X*-axis represents income values and the *Y*-axis represents the number of gold products. The plot is limited to X-axis limits of 0 to 100,000 and Y-axis limits of 0 to 100, with a customized size of 8 inches in width and 3 inches in height.

Breaking a DataFrame into batches

The `to_pandas_batches()` function converts a Snowpark DataFrame into multiple smaller pandas DataFrames to be processed in batches. This approach reduces memory usage by converting the data into manageable portions, enabling efficient analysis of large datasets:

```
for batch in marketing_final.to_pandas_batches(): print(batch.shape)
```

Here's the output:

```
(1748, 29)
(492, 29)
```

Figure 3.42 – DataFrame batches

The preceding code demonstrates how to analyze a large dataset in batches using the `to_pandas_batches()` function in Snowpark Python. By iterating over the `to_pandas_batches()` function, the code processes the dataset in manageable batches rather than loading the entire dataset into memory at once. In each iteration, a batch of the dataset is converted into a pandas DataFrame and stored in the `batch` variable. The `print(batch.shape)` statement provides the shape of each batch, indicating the number of rows and columns in that specific batch.

Analyzing the dataset in batches allows for more efficient memory utilization, enabling us to process large datasets that might otherwise exceed available memory resources. This approach facilitates the analysis of large datasets by breaking them into smaller, more manageable portions, allowing for faster computations and reducing the risk of memory-related issues.

Sampling a DataFrame

The `sample()` function in Snowpark Python allows us to retrieve a random subset of data from the Snowpark DataFrame. By specifying the desired fraction or number of rows, we can efficiently extract a representative sample for analysis. This technique reduces the memory footprint required for conversion and subsequent analysis while providing meaningful insights:

```
sample_df = marketing_final.sample(frac=0.50)
sample_df.count()
```

Here's the output:

```
1128
```

Figure 3.43 – Sampling data

The preceding code selects a random sample of 50% of the rows from the `marketing_final` DataFrame and assigns it to the `sample_df` DataFrame. The final count step produces slightly different output each time you run the code segment as it involves sampling the original table. The subsequent `sample_df.count()` function calculates the count of non-null values in each column of the `sample_df` DataFrame.

By utilizing the methods we covered here in Snowpark Python, we can overcome the limitations of converting large Snowpark DataFrames into pandas DataFrames, allowing for practical analysis while efficiently managing memory resources. These functions provide flexibility and control, enabling us to work with sizable datasets in a manageable and optimized manner.

Summary

Snowpark provides different data processing capabilities and supports various techniques. It provides us with an easy and versatile way to ingest different structured and unstructured file formats, and Snowpark's DataFrames support various data transformation and analysis operations. We covered various Snowpark session variables and different data operations that can be performed using Snowpark.

In the next chapter, we will cover how to build data engineering pipelines with Snowpark.

Building Data Engineering Pipelines with Snowpark

Data is the heartbeat of every organization, and data engineering is the lifeblood that ensures that current, accurate data is flowing through for various consumption. The role of a data engineer is to develop and manage the data engineering pipeline and the process that collects, transforms, and delivers data to a different **line of business (LOB)**. As Gartner's research rightly mentions, "*The increasing diversity of data, and the need to provide the right data to the right people at the right time, has created a demand for the data engineering practice. Data and analytics leaders must integrate the data engineering discipline into their data management strategy.*" This chapter discusses a practical approach to building efficient data engineering pipelines with Snowpark.

In this chapter, we're going to cover the following main topics:

- Developing resilient data pipelines with Snowpark
- Deploying efficient DataOps in Snowpark
- Overview of tasks in Snowflake
- Implementing logging and tracing in Snowpark

Technical requirements

This chapter requires an active Snowflake account and Python installed with Anaconda and configured locally. You can sign up for a Snowflake trial account at `https://signup.snowflake.com/`.

The technical requirements for environment setup are the same as in the previous chapters. If you haven't set up your environment yet, please refer to the previous chapter. Supporting materials are available at `https://github.com/PacktPublishing/The-Ultimate-Guide-To-Snowpark`.

Developing resilient data pipelines with Snowpark

A robust and resilient data pipeline will equip organizations to source, collect, analyze, and effectively use insights to grow business and deliver cost-saving business processes. Traditional data pipelines are difficult to manage and do not support the organization's evolving data needs. Snowpark solves problems that conventional data pipelines have by running them natively on the Snowflake Data Cloud and making extracting information from data in the Data Cloud easier and faster. This section will cover the various characteristics of resilient data pipelines, how to develop them in Snowpark, and their benefits.

Traditional versus modern data pipelines

A significant challenge of a traditional data pipeline is that it takes considerable time and cost to develop and manage, with high technical debt. It also consists of multiple tools that take much time to integrate. Due to the complexity of the solution, there is a possibility of delayed data due to latency and support for streaming data. The following diagram highlights the traditional method:

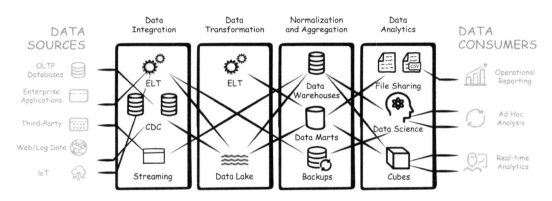

Figure 4.1 – Traditional data pipeline

The architecture shows a complex of technologies and systems being stitched together to deliver data from the source to consumers, with multiple points of failure at each stage. There are also issues of data governance and security due to data silos being present with numerous copies of the same data.

Modern data pipelines work based on a unified platform and multi-workload model. They integrate data sources such as batch and streaming to enhance productivity with streamlined architecture by enabling various **business intelligence (BI)** and analytics workloads and supporting internal and external users. The unified platform architecture supports continuous, extensible data processing pipelines with scalable performance. The following diagram highlights the modern Snowflake approach:

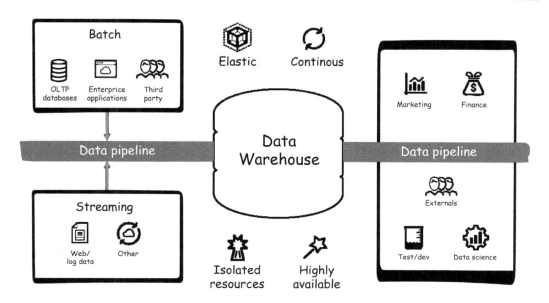

Figure 4.2 – Modern data pipeline

Snowpark stands out as a modern data pipeline tool due to its native integration with Snowflake, a leading cloud data warehouse, enabling seamless data processing directly within Spark applications. Offering a unified development experience with familiar programming languages such as Scala and Java, Snowpark eliminates the complexity associated with traditional Spark setups, allowing for streamlined development and maintenance of data pipelines. Snowpark's optimized performance for Snowflake's architecture ensures efficient data processing and reduced latency, enabling quick analysis of large datasets. Moreover, its advanced analytics capabilities, scalability, and cost-effectiveness make it a compelling choice for organizations seeking to build agile, cloud-native data pipelines with enhanced productivity and flexibility compared to traditional Spark setups.

Data engineering with Snowpark

Snowpark has many data engineering capabilities, making it a fast and flexible platform that enables developers to use Python for data engineering. With the support of **extract, transform, and load** (ETL) and **extract, load, and transform** (ELT), developers can use the Snowpark client for development and interact with the Snowflake engine for processing using their favorite developer environment. With the support of Anaconda, you can ensure that required packages and dependencies are readily available for Snowpark scripts. And it becomes easier to accelerate the growth of product pipelines. Data pipelines in Snowpark can be batch or real-time, utilizing scalable, high-performant multi-cluster warehouses capable of handling complex data transformations without compromising performance:

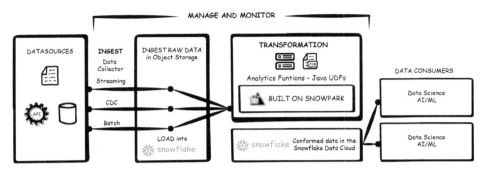

Figure 4.3 – Snowpark data engineering

Snowpark enhances the entire data engineering lifecycle with an engine that enables expressiveness and flexibility for developers with the simplicity of Data Cloud operations. Snowflake can help make data centralized, including structured, semi-structured, and unstructured data loaded into Snowflake for processing. Transformations can be carried out with the help of the powerful Python-based Snowpark functions. Snowpark data engineering workloads can be fully managed alongside other Snowflake objects throughout the development lifecycle with built-in monitoring and orchestration capabilities that support complex data pipelines at scale powered by the Data Cloud. The result of advanced data transformations is stored inside Snowflake and can be used for different data consumers. Snowpark data pipelines reduce the number of stages data needs to move to actionable insights by removing the step that moves data for computation.

Implementing programmatic ELT with Snowpark

Snowflake supports and recommends a modern ELT implementation pattern for data engineering instead of the legacy ETL process. ETL is a pattern where data is extracted from various sources, transformed in the data pipeline, and then the transformed data is loaded into a destination such as a data warehouse or data mart. The following diagram shows the comparison of ETL versus ELT:

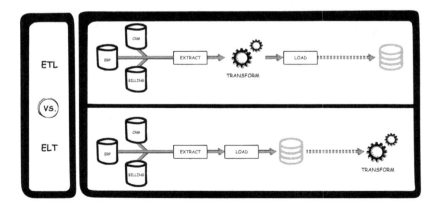

Figure 4.4 – ETL versus ELT

ELT is a pattern that is more suited for the Snowflake Data Cloud, where data is extracted from the source and loaded into Snowflake. This data is then transformed within Snowflake using Snowpark. Snowpark pipelines are designed to extract and load the data first and then transform it in the destination as the transformation is done inside Snowflake, which provides better scalability and elasticity. The ELT also improves performance and reduces the time it takes to ingest, transform, and analyze the data within Snowflake using Snowpark. The following diagram shows the different layers of data within Snowflake:

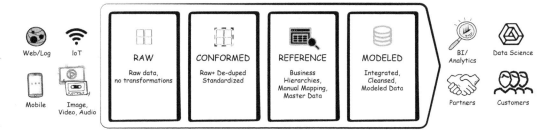

Figure 4.5 – Data stages in Snowflake

The data pipelines build the different data stages inside Snowflake. These stages are databases and schemas with objects such as tables and views inside them. The raw data is ingested from the source systems into Snowflake with no transformations since Snowflake supports multiple data formats. This data is then transformed using Snowpark into the conformed stage containing the de-duped and standardized data. This becomes the data that feeds into the next step in the data pipeline. The reference stage has the business definition and data mappings with the hierarchies and the master data. The final stage has the modeled data, which has the clean and transformed data. Snowpark has many functions that help with doing value-added transformations that help convert data into a business-ready format accessible to users and applications, making it more valuable for the organization.

ETL versus ELT in Snowpark

Snowpark supports both ETL and ELT workloads. While ELT is famous for modern pipelines, the ETL pattern is also used in some scenarios. ETL is commonly used with structured data where the total volume of data is small. It is also used in the system for migrating legacy databases to the Data Cloud where the source and target data types differ. ELT provides a significant advantage compared to the traditional ETL process. ELT supports large volumes of structured, unstructured, and semi-structured data that can be processed using Snowflake. It also allows developers and analysts to experiment with data as it is loaded into Snowflake. ELT also maximizes the option for them to transform data to get potential insights. It also supports low latency and real-time analytics. ELT is better suited for Snowflake than the traditional ETL for these reasons. The following section will cover how to develop efficient DataOps in Snowpark.

Deploying efficient DataOps in Snowpark

DataOps helps data teams reduce development times, increase data quality, and maximize the business value of data by bringing more rigor to the development and management of data pipelines. It also ensures that the data is clean, accurate, and up-to-date in a streamlined environment with data governance. Data engineering introduces the processes and capabilities required to effectively develop, manage, and deploy data engineering pipelines. The following diagram highlights the DataOps approach:

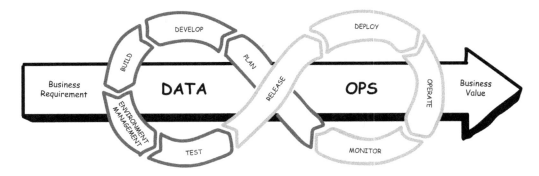

Figure 4.6 – DataOps process

The DataOps process focuses on bringing agile development to data engineering pipelines using an iterative development, testing, and deployment process in loops. It also includes **continuous integration and continuous deployment (CI/CD)** for data, schema changes, and the data versioning and automation of data models and artifacts. This section will show an example of a data engineering pipeline executed in Snowpark.

Developing a data engineering pipeline

Creating a resilient data engineering pipeline within the Snowpark framework requires integrating three core components seamlessly:

1. First and foremost, data engineers must master the art of loading data into Snowflake, setting the stage for subsequent processing. This initial step sets the foundation upon which the entire pipeline is built.

2. Second, the transformative power of Snowpark data functions comes into play, enabling engineers to shape and mold the data to meet specific analytical needs. *Chapter 3, Simplifying Data Processing Using Snowpark* provided a detailed exploration of DataFrame operations, laying the groundwork for this pivotal transformation phase.

3. Finally, the data journey culminates in bundling these operations as Snowpark stored procedures, offering efficiency and repeatability in handling data.

As we delve into this section, building upon the knowledge garnered from *Chapter 2, Establishing a Foundation with Snowpark* and *Chapter 3, Simplifying Data Processing Using Snowpark* where we elaborated on DataFrame operations and their conversion into **user-defined functions** (**UDFs**) and stored procedures, we will unravel the intricate process of unifying these elements into a resilient data engineering pipeline. This chapter is a testament to the synthesis of theory and practice, empowering data professionals to seamlessly interconnect the loading, transformation, and bundling phases, resulting in a robust framework for data processing and analysis within the Snowpark ecosystem.

With a comprehensive understanding of data loading from our discussions in the previous chapters, our journey now pivots toward strategically utilizing this data. This pivotal transition places our emphasis on three core steps:

1. Data preparation
2. Data transformation
3. Data cleanup

These stages constitute the cornerstone of our data engineering voyage, where we will sculpt, consolidate, and refine our data, revealing its true potential for analysis and valuable insights. We'll now transform these concepts into practical data engineering pipelines, leveraging the valuable insights from our prior discussions on stored procedure templates and transformation steps. Our focus will center on our marketing campaign data, where the foundational loading steps have been thoughtfully outlined in *Chapter 3, Simplifying Data Processing Using Snowpark* providing a solid starting point for our data preparation.

Data preparation

In the progression of our data engineering pipeline, the subsequent imperative phase is data preparation, which involves the integration of diverse tables. In this section, we will explore techniques for merging these disparate data tables using various functions tailored to the task. Additionally, we will elucidate the process of registering these functions as stored procedures, ensuring a streamlined and efficient data workflow. The first step is joining the purchase history with the campaign information. Both tables are entered using the `ID` column, and a single ID is retained:

```
def combine_campaign_table(purchase_history,campaign_info):
    purchase_campaign = purchase_history.join(
        campaign_info, \
        purchase_history.ID == campaign_info.ID, \
        lsuffix="_left", rsuffix="_right"
    )
    purchase_campaign = purchase_campaign.drop("ID_RIGHT")
    return purchase_campaign
```

The resultant `purchase_campaign` DataFrame holds the data and is used in the next step. In the next step, we join the purchase campaign with the complaint information using the same ID column and then create a `purchase_campaign_complain` DataFrame:

```
def combine_complain_table(purchase_campaign,complain_info):
    purchase_campaign_complain = purchase_campaign.join(
        complain_info, \
        purchase_campaign["ID_LEFT"] == complain_info.ID
    )
    purchase_campaign_complain = \
        purchase_campaign_complain.drop("ID_LEFT")
    return purchase_campaign_complain
```

The preceding code joins the column to create a `purchase_campaign_complain` DataFrame, which contains the mapped purchase data with complaint information. In the final step, a marketing table is created by the union of the data between the purchase complaint and the marketing table:

```
def union_marketing_additional_table(
    purchase_campaign_complain,marketing_additional):
    final_marketing_table = \
        purchase_campaign_complain.union_by_name(
            marketing_additional
        )
    return final_marketing_table
```

The preceding code produces a table that contains all the combined data that is the final result of the pipeline, which will be written as a table. The Python functions representing each step are executed as part of the Snowpark stored procedure. The stored procedures can be performed in sequence one after the other and also scheduled as Snowflake tasks. The data preparation procedure calls the three Python methods, and the final table is written to Snowflake:

```
from snowflake.snowpark.functions import sproc
import snowflake

def data_prep(session: Session):

    #### Loading Required Tables
    purchase_history = session.table("PURCHASE_HISTORY")
    campaign_info = session.table("CAMPAIGN_INFO")
    complain_info = session.table("COMPLAINT_INFO")
    marketing_additional = session.table("MARKETING_ADDITIONAL")
```

The preceding code does the data preparation by loading the required data into the DataFrame. We will now call each of the steps to execute it like a pipeline:

```
#### Calling Step 1
purchase_campaign = combine_campaign_table(
    purchase_history, campaign_info)
#### Calling Step 2
purchase_campaign_complain = combine_campaign_table(
    purchase_campaign, complain_info)
#### Calling Step 3
final_marketing_data = union_marketing_additional_table(
    purchase_campaign_complain, marketing_additional)
```

The three previously defined step functions are executed. The resultant data is loaded into the new `final_marketing_data` DataFrame, which will then be loaded to the Snowflake table:

```
#### Writing Combined Data To New Table
final_marketing_data.write.save_as_table( \
    "FINAL_MARKETING_DATA")
return "LOADED FINAL MARKETING DATA TABLE"
```

Now, we will create and execute a stored procedure that contains the preceding logic. The procedure is called `data_prep_sproc` and is the first part of the data engineering pipeline – data preparation:

```
# Create an instance of StoredProcedure using the sproc() function
from snowflake.snowpark.types import IntegerType,StringType
data_prep_sproc = sproc(
                       func= data_prep,\
                       replace=True,\
                       return_type = StringType(),\
                       stage_location="@my_stage",\
                       packages=["snowflake-snowpark-python"]
                       )
```

The preceding stored procedure writes the data into the `Final_Marketing_Data` table, which will be used in the next step of data transformation.

Data transformation

The subsequent phase in this process involves data transformation, building upon the data prepared in the previous step. Here, we'll take the pivotal action of registering another stored procedure after the last stage. This procedure applies transformation logic, molding the data into a form primed for analysis. Leveraging Snowpark's array of valuable aggregation and summarization functions, we will harness these capabilities to shape and enhance our data, laying a solid foundation for rigorous analysis. The following code transforms the data:

```
def data_transform(session: Session):

    #### Loading Required Tables
    marketing_final = session.table("FINAL_MARKETING_DATA")
    market_subset = marketing_final.select("EDUCATION", \
        "MARITAL_STATUS","INCOME")
    market_pivot = market_subset.pivot("EDUCATION", \
        ["Graduation","PhD","Master","Basic","2n Cycle"]
    ).sum("INCOME")

    #### Writing Transformed Data To New Table
    market_pivot.write.save_as_table("MARKETING_PIVOT")
    return "CREATED MARKETING PIVOT TABLE"

data_transform_sproc = sproc(
                    func= data_transform,\
                    replace=True,\
                    return_type = StringType(),\
                    stage_location="@my_stage",\
                    packages=["snowflake-snowpark-python"]
                    )
```

A data_transform_sproc stored procedure is created, which reads the Final_Marketing_Data table and creates a pivot with the education of the customer and the total income. When the stored procedure is executed, this is then written to the Marketing_Pivot table.

Data cleanup

In the final step of our data engineering process, we focus on a crucial task: cleaning up the data in the Marketing_Pivot table. Similar to artists perfecting a masterpiece, we carefully go through our data, removing any empty values in tables that aren't important for our analysis. To do this, we rely on the versatile dropna() function, which acts like a precise tool to cut away unnecessary data:

```
def data_cleanup(session: Session):
```

```
#### Loading Required Tables
market_pivot = session.table("MARKETING_PIVOT")

market_drop_null = market_pivot.dropna(thresh=5)

#### Writing Cleaned Data To New Table
market_drop_null.write.save_as_table("MARKET_PIVOT_CLEANED")
return "CREATED CLEANED TABLE"

data_cleanup_sproc = sproc(
                    func= data_cleanup,\
                    replace=True,\
                    return_type = StringType(),\
                    stage_location="@my_stage",\
                    packages=["snowflake-snowpark-python"]
                    )
```

The cleaned-up data from the market_drop_null DataFrame is then saved into the Market_Pivot_Cleaned table. This data is at the last stage of the pipeline and is used for analysis.

Orchestrating the pipeline

The data pipeline is orchestrated by calling three Snowpark procedures, which invokes the three different steps of the data engineering pipeline. The procedures are executed in the order of data_prep_sproc, data_transform_sproc, and data_cleanup_sproc:

```
#### Calling Data Preparation Stored Procedure
data_prep_sproc()

#### Calling Data Transformation Stored Procedure
data_transform_sproc()

#### Calling Data Cleanup Stored Procedure
data_cleanup_sproc()
```

The Snowpark procedure is executed, and after each step is executed, the final data is written to the Market_Pivot_Cleaned table. Snowflake supports scheduling and orchestration through tasks. Tasks can be scheduled using the Python API and through worksheets, and they can trigger procedures in sequence:

Figure 4.7 – Stored procedure execution

In the following section, we will explore how we can utilize Snowflake tasks and task graphs to execute the preceding pipeline.

Overview of tasks in Snowflake

Tasks in Snowflake are powerful tools designed to streamline data processing workflows and automate various tasks within the Snowflake environment. Offering a range of functionalities, tasks execute different types of SQL code, enabling users to perform diverse operations on their data.

Tasks in Snowflake can execute three main types of SQL code:

- **Single SQL statement**: Allows the execution of a single SQL statement
- **Call to a stored procedure**: Enables the invocation of a stored procedure
- **Procedural logic using Snowflake Scripting**: Supports the implementation of procedural logic using Snowflake Scripting

Tasks can be integrated with table streams to create continuous ELT workflows. By processing recently changed table rows, tasks ensure the maintenance of data integrity and provide exactly-once semantics for new or altered data. Tasks in Snowflake can be scheduled to run at specified intervals. Snowflake ensures that only one instance of a scheduled task is executed at a time, skipping scheduled executions if a task is still running.

Compute models for tasks

In the serverless compute model, tasks rely on compute resources managed by Snowflake. These resources are automatically resized and scaled based on workload demands, ensuring optimal performance and resource utilization. Snowflake dynamically determines the appropriate compute size for each task run based on historical statistics.

Alternatively, users can opt for the user-managed virtual warehouse model, where they specify an existing virtual warehouse for individual tasks. This model provides users with more control over compute resource management but requires careful sizing to ensure efficient task execution.

Task graphs

Task graphs, also known as **directed acyclic graphs (DAGs)**, allow for the organization of tasks based on dependencies. Each task within a task graph has predecessor and subsequent tasks, facilitating complex workflow management.

Task graphs are subject to certain limitations, including a maximum of 1,000 tasks in total, including the root task. Individual tasks within a task graph can have a maximum of 100 predecessors and 100 child tasks.

Users can view and monitor their task graphs using SQL or Snowsight, Snowflake's integrated development environment, providing visibility into task dependencies and execution status.

In summary, tasks in Snowflake offer robust capabilities for data processing, automation, and workflow management, making them indispensable tools for users seeking to optimize their data operations within the Snowflake ecosystem.

Managing tasks and task graphs with Python

Our primary focus is on Snowpark. Now, we'll explore how we can utilize Python Snowpark to programmatically perform task graph operations instead of using SQL statements.

Now, Python can manage Snowflake tasks, allowing users to run SQL statements, procedure calls, and Snowflake Scripting logic. The Snowflake Python API introduces two types:

- `Task`: This type represents a task's properties, such as its schedule, parameters, and dependencies
- `TaskResource`: This type provides methods to interact with `Task` objects, enabling task execution and modification

Tasks can be grouped into task graphs, which consist of interconnected tasks arranged based on their dependencies. To create a task graph, users first define a DAG object, specifying its name and optional properties, such as its schedule. The scheduling of a task graph can be customized using either a `timedelta` value or a cron expression, allowing for flexible task execution timing and recurrence patterns.

Let's begin by setting up the necessary functions to implement our DAG. The examples provided in this section presuppose that you've already written code to establish a connection with Snowflake to utilize the Snowflake Python API:

```
from snowflake.core import Root
from snowflake.core.task import StoredProcedureCall
from snowflake.core.task.dagv1 import DAG, DAGTask, DAGOperation
from snowflake.snowpark import Session
from datetime import timedelta
root = Root(session)
```

The preceding code initializes the Snowflake Python API, creating a `root` object for utilizing its types and methods. Additionally, it sets up a `timedelta` value of 1 hour for the task's schedule. You can define the schedule using either a `timedelta` value or a Cron expression. For one-off runs, you can omit the schedule argument to the DAG object without worrying about it running unnecessarily in the background.

Let's define a simple DAG that we'll use to execute our pipelines:

```
dag = DAG("Task_Demo",
          warehouse="COMPUTE_WH",
          schedule=timedelta(days=1),
          stage_location= \
              "SNOWPARK_DEFINITIVE_GUIDE.MY_SCHEMA.MY_STAGE",
          packages=["snowflake-snowpark-python"]
          )
```

In this DAG setup, the following applies:

- We've named our DAG Task_Demo, which by default runs on the specified warehouse.

- A schedule for daily execution has been defined using timedelta.

- A stage_location attribute is necessary for storing the serialized version of the tasks via the Python API.

- All tasks under this DAG will run with the default list of packages and the specified warehouse. However, both the warehouse and packages for individual tasks within the DAG can be overridden with different values.

- In addition, the use_func_return_value attribute indicates that the return value of Python functions will be treated as the return value of the task, though in our case, we're not utilizing the return object.

We've now defined a series of Python functions representing a three-task pipeline, or DAG. However, we haven't yet created and pushed the DAG to Snowflake. Let's do that now using the Snowflake Python API:

```
with dag:
    data_prep_task = DAGTask("Data_Prep", definition=data_prep)
    data_transform_task = DAGTask("Data_Transform", \
        definition=data_transform)
    data_cleanup_task = DAGTask("Data_Cleanup", \
        definition=data_cleanup)
    data_prep_task >> data_transform_task >> data_cleanup_task
```

In this code snippet, we've instantiated DAGTask objects for each task in our pipeline: data_prep, data_transform, and data_cleanup. These tasks are then linked together using the >> operator to specify their execution order.

For one-off testing or running of the DAG, users can skip specifying the schedule and manually trigger a run with dag_op.run(dag):

```
schema = root.databases["SNOWPARK_DEFINITIVE_GUIDE"].schemas[ \
    "MY_SCHEMA"]
dag_op = DAGOperation(schema)
dag_op.deploy(dag,mode="orReplace")
dag_op.run(dag)
```

The provided code performs several actions using Snowflake's Snowpark library. Firstly, it retrieves the schema named MY_SCHEMA from the SNOWPARK_DEFINITIVE_GUIDE database using the root object. Then, it initializes a DAGOperation object named dag_op, specifying the schema where the DAG will be deployed. The deploy() method is then called on dag_op to deploy the specified DAG (named dag) in the specified schema.

The orReplace mode argument indicates that if a DAG with the same name already exists in the schema, it will be replaced. Finally, the run() method is called on dag_op to execute the deployed DAG. This code essentially sets up and executes a DAG within Snowflake using Snowpark. Now, you can check the graph in Snowsight to see how the graph has been set up:

Figure 4.8 – Snowsight graph

Additionally, note that you can pass the raw function to the dependency definition without explicitly creating a DAGTask instance, with the library automatically creating a task for you with the same name. However, there are exceptions to this rule, such as needing to explicitly create a DAGTask instance for the first task or when utilizing DAGTaskBranch or repeating certain functions in multiple tasks.

Once you've deployed and run the DAG, you can easily check its status using the following code:

```
current_runs = dag_op.get_current_dag_runs(dag)
for r in current_runs:
    print(f"RunId={r.run_id} State={r.state}")
```

As depicted in the screenshot, our tasks are scheduled to run daily. Additionally, we have executed them once to verify the task deployment beforehand:

RunId=1712744391720 State=SCHEDULED

Figure 4.9 – Tasks deployed

As we progress in constructing more complex pipelines, managing and debugging become increasingly challenging. It's crucial to establish a robust logging mechanism to facilitate maintenance and streamline error resolution. In the next section, we'll delve into implementing logging and traceback functionalities in Snowpark to enhance our pipeline's manageability and ease troubleshooting.

Implementing logging and tracing in Snowpark

Logging and tracing are crucial for DataOps and are necessary to monitor and fix failures in the data engineering pipeline. Snowpark comes with logging and tracing functionality that is built in, which can help record the activity of Snowpark functions and procedures and capture those in an easy-to-access central table inside Snowflake. Log messages are independent, detailed messages with information in the form of strings, providing details about the piece of code, and trace events are structured data that we can use to get information spanning and grouping multiple parts of our code. Once logs are collected, they can be easily queried by SQL or accessed via Snowpark. The following diagram highlights the event table and alerting:

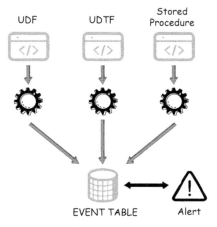

Figure 4.10 – Event table

Snowpark stores logs and trace messages inside the event table, a unique table with a predefined set of columns. Logs and traces are captured in this table as the code is executed. Let's look at the structure of event tables and how to create them.

Event tables

An event table is native to Snowflake and needs to be created. There can be only one event table for a Snowflake account that captures all the information, but multiple views can be made for analysis. An event table contains the following columns:

Column	Data Type	Description
TIMESTAMP	TIMESTAMP_NTZ	The UTC timestamp when an event was created. This is the end of the period for events representing a period.
START_TIMESTAMP	TIMESTAMP_NTZ	For events representing a period, such as trace events as the start of the period as a UTC timestamp.
OBSERVED_TIMESTAMP	TIMESTAMP_NTZ	A UTC timestamp is used for logs. Currently, it has the same value as TIMESTAMP.
TRACE	OBJECT	Tracing context for all signal types. Contains trace_id and span_id string values.
RESOURCE	OBJECT	Reserved for future use.
RESOURCE_ATTRIBUTES	OBJECT	Attributes that identify the source of an event, such as database, schema, user, warehouse, and so on.
SCOPE	OBJECT	Scopes for events; for example, class names for logs.
SCOPE_ATTRIBUTES	OBJECT	Reserved for future use.
RECORD_TYPE	STRING	The event types. One of the following: LOG for a log message. SPAN for UDF invocations performed sequentially on the same thread. SPAN_EVENT for a single trace event. A single query can emit more than one SPAN_EVENT event type.
RECORD	OBJECT	Fixed values for each record type.

RECORD_ATTRIBUTES	OBJECT	Variable attributes for each record type.
VALUE	VARIANT	Primary event value.
EXEMPLARS	ARRAY	Reserved for future use.

Table 4.1 – Event table columns

Each column can be queried or combined to analyze different outcomes based on the logs and the traces. The log type describes the log levels assigned as part of the logging and can be set at both objects and the session. The log levels can be the following:

LOG_LEVEL Parameter Setting	Levels of Messages Ingested
TRACE	TRACE
	DEBUG
	INFO
	WARN
	ERROR
	FATAL
DEBUG	DEBUG
	INFO
	WARN
	ERROR
	FATAL
INFO	INFO
	WARN
	ERROR
	FATAL
WARN	WARN
	ERROR
	FATAL
ERROR	ERROR
	FATAL
FATAL	ERROR
	FATAL

Table 4.2 – Event table log levels

The log level is a hierarchy applied in the order presented in *Table 4.2*. In the next section, we will look at creating an event table.

> **Note**
>
> It is best practice to set the necessary log level based on a minor level such as FATAL and ERROR so that the number of logged messages is fewer. In the case of logging INFO, it is usually turned on to capture the log and turned off in production to avoid catching too many records.

Creating and configuring an event table

The first step is to create an event table for the Snowflake account. The name of the event table can be specified, and columns for the event table are not required to be set when creating the table, as Snowflake automatically creates it with the standard columns. The event table is assigned to the account and needs to be made in a separate database that does not have Snowflake replication enabled using the ACCOUNTADMIN role.

To create an event table, run the following command with the ACCOUNTADMIN role using Snowpark:

```
session.sql('''CREATE EVENT TABLE MY_EVENTS;''').show()
```

An event table called MY_EVENTS is created with the default column structure. The next step is to assign the event table as the active event table to a particular Snowflake account. The event table can be assigned to an account by executing the following code with the ACCOUNTADMIN role:

```
session.sql('''ALTER ACCOUNT SET EVENT_TABLE = \
    SNOWPARK_DEFINITIVE_GUIDE.MY_SCHEMA.MY_EVENTS;
''').show()
```

The parameter is applied at the account level, and all events from the particular Snowflake account are captured in this event table. This completes the event table setup.

Querying event tables

An event table can be accessed just like any other Snowflake table. To get the records from an event table, you can query the event table with the following code:

```
session.sql('''SELECT *
    FROM SNOWPARK_DEFINITIVE_GUIDE.MY_SCHEMA.MY_EVENTS;
''').show()
```

It returns a result with empty records since no information was captured. Records in an event table can be filtered with a specific column to get detailed information. A Snowflake stream can be set on top of the event table to capture only new events. A Stream can be created by running the following query:

```
session.sql('''CREATE STREAM EVENT_APPEND ON EVENT TABLE MY_EVENTS
APPEND_ONLY=TRUE;''').show()
```

The EVENT_APPEND stream captures the latest inserted records into the event table. In the next section, we will set up logging and tracing to capture records in an event table.

Setting up logging in Snowpark

Introducing logging and tracing capabilities into our pipelines is akin to infusing resilience into our standard data engineering processes. This section will delve into the integration of logging functionalities into our existing data engineering pipeline. By doing so, we not only gain the ability to monitor and trace the flow of data but also enhance the robustness of our code, fortifying it against potential pitfalls. Join us as we explore how these logging capabilities elevate our data engineering practices, making them more reliable and fault-tolerant.

Logging can be enabled for both Snowpark functions and procedures. The first step in capturing logs is to set the log level in the Snowpark session. The log level in the session can be set by running the following code:

```
session.sql('''alter session set log_level = INFO;''').show()
```

This sets the log level to INFO for the particular session, so all Snowpark execution that happens for the specific session is captured with the log level as information. Snowpark supports APIs to log messages directly from the handler.

Capturing informational logs

We will now modify the data preparation procedures from the data pipeline to capture informational logs. We start with prepping the data:

```
from snowflake.snowpark.functions import sproc
import logging

def data_prep(session: Session):

    ## Initializing Logger
    logger = logging.getLogger("My_Logger")
    logger.info("Data Preparation Pipeline Starts")

    #### Loading Required Tables
    logger.info("Loading Required Tables")
```

```
purchase_history = session.table("PURCHASE_HISTORY")
campaign_info = session.table("CAMPAIGN_INFO")
complain_info = session.table("COMPLAINT_INFO")
marketing_additional = session.table("MARKETING_ADDITIONAL")
```

The data from the four tables is loaded into the DataFrame. This DataFrame is then used to execute each step. Next, we will proceed with calling each step in order to process the data:

```
#### Calling Step 1
purchase_campaign = combine_campaign_table(
    purchase_history,campaign_info)

logger.info("Joined Purchase and Campaign Tables")

#### Calling Step 2
purchase_campaign_complain = combine_complain_table(
    purchase_campaign,complain_info)

logger.info("Joined Complain Table")

#### Calling Step 3
final_marketing_data = union_marketing_additional_table(
    purchase_campaign_complain,marketing_additional)

logger.info("Final Marketing Data Created")
```

Once all three steps are executed, we get the final marketing data ready to be loaded into a Snowflake table for consumption. The following code will load the data into a Snowflake table:

```
#### Writing Combined Data To New Table
final_marketing_data.write.save_as_table( \
    "FINAL_MARKETING_DATA")

logger.info("Final Marketing Data Table Created")
return "LOADED FINAL MARKETING DATA TABLE"
```

The data is loaded into a table called FINAL_MARKETING_DATA. The table is automatically created with the data in the Snowpark DataFrame. We will now register this as a Snowpark stored procedure:

```
## Register Stored Procedure in Snowflake
### Add packages and data types
from snowflake.snowpark.types import StringType
session.add_packages('snowflake-snowpark-python')
```

```
### Upload Stored Procedure to Snowflake
session.sproc.register(
    func = data_prep
  , return_type = StringType()
  , input_types = []
  , is_permanent = True
  , name = 'DATA_PREP_SPROC_LOG'
  , replace = True
  , stage_location = '@MY_STAGE'
)
```

The logging module from Python's standard library is used for logging. The package is imported, and the logger's name is specified as My_Logger. Different logger names can be set for processes to help identify the particular logging application. We can execute the stored procedure to capture logs to the event table. The stored procedure can be executed by running the following command:

```
session.sql(''' Call DATA_PREP_SPROC_LOG()''').show()
```

The stored procedure is executed, and we can see the successful output of the execution:

Figure 4.11 – Stored procedure execution

The following section will cover how to query logs generated by the stored procedure.

Querying informational logs

Logs generated by the stored procedure can be accessed from the event table. Records captured from the previous stored procedure can be accessed by running the following query:

```
session.sql("""
    SELECT RECORD['severity_text'] AS SEVERITY,
        VALUE AS MESSAGE
    FROM MY_EVENTS
    WHERE SCOPE['name'] = 'My_Logger'
    AND RECORD_TYPE = 'LOG'
""").show()
```

We filter logs captured only from My_Logger by specifying it in the scope. The query returns the following records that were generated from executing the stored procedure:

```
|"SEVERITY"    |"MESSAGE"                              |

|"INFO"        |"Data Preparation Pipeline Starts"    |
|"INFO"        |"Loading Required Tables"             |
|"INFO"        |"Joined Purchase and Campaign Tables" |
|"INFO"        |"Joined Complain Table"               |
|"INFO"        |"Final Marketing Data Created"        |
|"INFO"        |"Final Marketing Data Table Created"  |
```

Figure 4.12 – Querying logs

In the next section, we will set up logging to capture error messages and handle exceptions in Snowpark.

Handling exceptions in Snowpark

Exception handling is integral to data pipelines as it helps identify and handle issues. Exception handling can be done by catching exceptions thrown from within the try block and capturing those error logs. The ERROR and WARN log levels are often used when capturing exceptions, and fatal issues are logged at the FATAL level. This section will look at capturing error logs and handling the exception on the pipeline.

Capturing error logs

We will modify the data transformation stored procedure to capture error logs and add exception handling:

```
def data_transform(session: Session):
    try:

        ## Initializing Logger
        logger = logging.getLogger("Data_Transform_Logger")
        logger.info("Data Transformation Pipeline Starts")

        ## Pivoting Process
        marketing_final = session.table("FINAL_MARKETING_DATA")
        market_subset = marketing_final.select("EDUCATION", \
            "MARITAL_STATUS","INCOME")
        market_pivot = market_subset.pivot("EDUCATION", \
            ["Graduation","PhD","Master","Basic","2n Cycle"]
        ).sum("INCOME")

        #### Writing Transformed Data To New Table
        market_pivot.write.save_as_table("MAREKTING_PIVOT")
        logger.log("MARKETING PIVOT TABLE CREATED")
        return "CREATED MARKETING PIVOT TABLE"
```

```
except Exception as err:
    logger.error("Logging an error from Python handler: ")
    logger.error(err)
    return "ERROR"
```

The data transformation logic is moved into the try block, and a logger named Data_Transform_Logger is initiated. The exception raised by the code is captured in the exception object defined as err. This is then logged in the event table by the ERROR log level. We will now register this stored procedure in Snowpark:

```
## Register Stored Procedure in Snowflake

### Add packages and data types
from snowflake.snowpark.types import StringType
session.add_packages('snowflake-snowpark-python')

### Upload Stored Procedure to Snowflake
session.sproc.register(
    func = data_transform
  , return_type = StringType()
  , input_types = []
  , is_permanent = True
  , name = 'DATA_TRANSFORM_SPROC_LOG_ERROR'
  , replace = True
  , stage_locations = "@MY_STAGE"
)
```

We will execute the stored procedure purposely with the error to log the error. The stored procedure can be triggered by running the following command:

```
session.sql(''' Call DATA_TRANSFORM_SPROC_LOG_ERROR()''').show()
```

The stored procedure has raised an exception, and the error has been captured in the event table:

Figure 4.13 – Error execution

The following section will cover querying error logs generated by the procedure.

Querying error logs

The error log records from the preceding stored procedure are captured under the Data_Transform_Logger logger, which can be accessed by filtering the query to return logs specific to the logger. The following query can be executed to get the records from the event table:

```
session.sql("""
    SELECT RECORD['severity_text'] AS SEVERITY,VALUE AS MESSAGE
    FROM MY_EVENTS
    WHERE SCOPE['name'] = 'Data_Transform_Logger'
    AND RECORD_TYPE = 'LOG'
""").collect()
```

The scope name is filtered as the Data_Transform_Logger to get the results, and errors caused by the exception are logged under SEVERITY as ERROR:

```
[Row(SEVERITY='"INFO"', MESSAGE='"Data Preparation Pipeline Starts"'),
 Row(SEVERITY='"ERROR"', MESSAGE='"Logging an error from Python handler: "'),
 Row(SEVERITY='"ERROR"', MESSAGE='"(1304): 01aea100-0001-0458-0001-9ba600015746: 002002 (42710): 01aea100-0001-0458-0001-9ba60
 Row(SEVERITY='"INFO"', MESSAGE='"Data Transformation Pipeline Starts"'),
 Row(SEVERITY='"ERROR"', MESSAGE='"Logging an error from Python handler: "'),
 Row(SEVERITY='"ERROR"', MESSAGE='"(1304): 01aea101-0001-0458-0001-9ba60001576a: 002002 (42710): 01aea101-0001-0458-0001-9ba60
```

Figure 4.14 – Error log messages

Snowpark makes debugging easy by supporting logging errors and handling exceptions through event tables. The following section will cover event traces and how to capture trace information in Snowpark.

Setting up tracing in Snowpark

Trace events are a type of telemetry data that is captured when something has happened in the code. It has a structured payload that helps analyze the trace by aggregating this information to understand the code's behavior at a high level. When a procedure or function executes, trace events are emitted, which are available in the active event table.

The first step in capturing events is to turn on the tracing functionality by setting the trace level in the Snowpark session. The trace level in the session can be set by running the following code:

```
session.sql("ALTER SESSION SET TRACE_LEVEL = ALWAYS;").show()
```

In the next section, we will look at capturing traces.

Capturing traces

Traces can be captured with the open source Snowflake telemetry Python package available in the Anaconda Snowflake channel. The package needs to be imported into the code, and it will be executed in Snowpark. The telemetry package can be imported by including the code in the Snowpark handler:

```
from snowflake import telemetry
```

The `telemetry` package helped capture traces generated on the code and logged into the event table. We will modify the data cleanup procedure by adding telemetry events to capture the trace:

```
def data_cleanup(session: Session):

    #### Loading Telemetry Package
    from snowflake import telemetry

    #### Loading Required Tables
    market_pivot = session.table("MARKETING_PIVOT")

    #### Adding Trace Event
    telemetry.add_event("data_cleanup", \
        {"table_name": "MARKETING_PIVOT", \
         "count": market_pivot.count()})

    #### Dropping Null
    market_drop_null = market_pivot.dropna(thresh=5)

    #### Writing Cleaned Data To New Table
    market_drop_null.write.save_as_table("MARKET_PIVOT_CLEANED")

    #### Adding Trace Event
    telemetry.add_event("data_cleanup", \
        {"table_name": "MARKET_PIVOT_CLEANED", \
         "count": market_drop_null.count()})

    return "CREATED CLEANED TABLE"
############################################################
## Register Stored Procedure in Snowflake

### Add packages and data types
from snowflake.snowpark.types import StringType
session.add_packages('snowflake-snowpark-python', \
    'snowflake-telemetry-python')

### Upload Stored Procedure to Snowflake
session.sproc.register(
    func = data_cleanup
  , return_type = StringType()
  , input_types = []
  , is_permanent = True
  , name = 'DATA_CLEANUP_SPROC_TRACE'
  , replace = True
```

```
    , stage_location = '@MY_STAGE'
)
```

The procedure is now ready to be executed. We are passing attributes for the operations captured on the traces. We can execute the procedure by running the following code:

```
session.sql(''' Call DATA_CLEANUP_SPROC_TRACE()''').show()
```

The procedure is executed, and the data is cleaned from the table:

Figure 4.15 – Data cleanup procedure execution

The traces are now generated in the event table. We can directly query the event table to obtain trace information.

Querying traces

Traces generated can be accessed from the event table. Traces captured from the previous stored procedure can be accessed by running the following query:

```
session.sql("""
    SELECT
        TIMESTAMP as time,
        RESOURCE_ATTRIBUTES['snow.executable.name']
            as handler_name,
        RESOURCE_ATTRIBUTES['snow.executable.type']
            as handler_type,
        RECORD['name'] as event_name,
        RECORD_ATTRIBUTES as attributes
    FROM
        MY_EVENTS
    WHERE
        EVENT_NAME ='data_cleanup'
""").show(2)
```

We are querying using the `data_cleanup` event name. This returns the two traces captured when the code was executed:

```
| "HANDLER_NAME"                              | "HANDLER_TYPE" | "EVENT_NAME"   | "ATTRIBUTES"                             |
| "DATA_CLEANUP_SPROC_TRACE():VARCHAR(16777216)" | "PROCEDURE"    | "data_cleanup" | {                                        |
|                                             |                |                |   "count": 8,                            |
|                                             |                |                |   "table_name": "MARKETING_PIVOT"        |
|                                             |                |                | }                                        |
| "DATA_CLEANUP_SPROC_TRACE():VARCHAR(16777216)" | "PROCEDURE"    | "data_cleanup" | {                                        |
|                                             |                |                |   "count": 5,                            |
|                                             |                |                |   "table_name": "MARKET_PIVOT_CLEANED"   |
|                                             |                |                | }                                        |
```

Figure 4.16 – Trace capture information

We can see the attributes that are captured from the execution of the stored procedure. After the NULL values have been cleaned up, the details return the total data counts, including the NULL and count values.

Comparison of logs and traces

The following table compares logs and traces and lists scenarios to use each:

Characteristic	Log entries	Trace events
Intended use	Record detailed but unstructured information about the state of your code. Use this information to understand what happened during a particular invocation of your function or procedure.	Record a brief but structured summary of each invocation of your code. Aggregate this information to understand the behavior of your code at a high level.
Structure as a payload	None. A log entry is just a string.	Structured with attributes you can attach to trace events. Attributes are key-value pairs that can be easily queried with a SQL query.
Supports grouping	No. Each log entry is an independent event.	Yes. Trace events are organized into spans. A span can have its attributes.
Quantity limits	Unlimited. All log entries emitted by your code are ingested into the event table.	The number of trace events per span is capped at 128. There is also a limit on the number of span attributes.
Complexity of queries against recorded data	Relatively high. Your queries must parse each log entry to extract meaningful information from it.	Relatively low. Your queries can take advantage of the structured nature of trace events.

Table 4.3 – Differences between logs and traces

Logs and traces help debug Snowpark and are a practical feature for efficient DataOps.

Summary

In this chapter, we covered in detail building and deploying resilient data pipelines in Snowpark and also how to enable logging and tracking using event tables. Building resilient data pipelines with effective DataOps is vital for a successful data strategy. Snowpark supports the development of a modern data pipeline through a programmatic ELT approach along with features such as logging and tracing, making it easy for developers to implement DataOps. We also covered how tasks and task graphs can be used in scheduling and deploying pipelines.

In the next chapter, we will cover using Snowpark to develop data science workloads.

5

Developing Data Science Projects with Snowpark

The emergence of cloud technologies has ushered in a new era of possibilities. With the advent of Data Cloud, a robust platform that unifies data storage, processing, and analysis, data scientists have many opportunities to explore, analyze, and extract meaningful insights from vast datasets. In this intricate digital realm, the role of Snowpark, a cutting-edge data processing framework, becomes paramount. This chapter serves as an illuminating guide, delving deep into developing data science projects with Snowpark, unraveling its intricacies, and harnessing its potential to the fullest extent.

In this chapter, we're going to cover the following main topics:

- Data science in Data Cloud
- Exploring and preparing data
- Training **machine learning** (ML) models in Snowpark

Technical requirements

For this chapter, you'll require an active Snowflake account and Python installed with Anaconda configured locally. You can sign up for a Snowflake Trial account at `https://signup.snowflake.com/`.

To configure Anaconda, follow `https://conda.io/projects/conda/en/latest/user-guide/getting-started.html`.

In addition, to install and set up Python for VS Code, follow `https://code.visualstudio.com/docs/python/python-tutorial`.

To learn how to operate Jupyter Notebook in VS Code, go to `https://code.visualstudio.com/docs/datascience/jupyter-notebooks`.

The supporting materials for this chapter are available in this book's GitHub repository at `https://github.com/PacktPublishing/The-Ultimate-Guide-To-Snowpark`.

Data science in Data Cloud

Data science transcends traditional boundaries in the Data Cloud ecosystem, offering a dynamic environment where data scientists can harness the power of distributed computing and advanced analytics. With the ability to seamlessly integrate various data sources, including structured and unstructured data, Data Cloud provides a data exploration and experimentation environment. We will start this section with a brief data science and ML refresher.

Data science and ML concepts

Data science and ML have surged to the forefront of technological and business innovation, becoming integral components of decision-making, strategic planning, and product development across virtually all industries. The journey to their current popularity and influence is a testament to several factors, including advancements in technology, the explosion of data, and the increasing computational power available. This section will briefly discuss data science and ML concepts.

Data science

Data science is a multidisciplinary field that relies on various software tools, algorithms, and ML principles to extract valuable insights from extensive datasets. Data scientists are pivotal in collecting, transforming, and converting data into predictive and prescriptive insights. By employing sophisticated techniques, data science uncovers hidden patterns and meaningful correlations within data, enabling businesses to act on informed decisions based on empirical evidence.

Artificial intelligence

Artificial intelligence (AI) encompasses the science and engineering of creating intelligent machines and brilliant computer programs capable of autonomously processing information and generating outcomes. AI systems are designed to solve intricate problems using logic and reasoning, similar to human cognitive processes. These systems operate autonomously, aiming to emulate human-like intelligence in decision-making and problem-solving tasks.

ML

ML, a subset of AI, involves specialized algorithms integrated into the data science workflow. These algorithms are meticulously crafted software programs that are designed to detect patterns, identify correlations, and pinpoint anomalies within data. ML algorithms excel at predicting outcomes based on existing data and continue to learn and improve their accuracy as they encounter new data and situations. Unlike humans, ML algorithms can process thousands of attributes and features, enabling the discovery of unique combinations and correlations in vast datasets. This capability makes ML indispensable for extracting valuable insights and predictions from extensive data collections.

Now that we have the terminologies straightened out, we will discuss how the Data Cloud paradigm has helped the growth of data science and ML for organizations.

The Data Cloud paradigm

Data science in the cloud represents a paradigm shift in how data-driven insights are derived and applied. In this innovative approach, data science processes, tools, and techniques are seamlessly integrated into the cloud, allowing organizations to leverage the power of scalable infrastructure and advanced analytics.

At the heart of this paradigm lies Data Cloud, a dynamic ecosystem that transcends traditional data storage and processing constraints. The Data Cloud paradigm represents a seismic shift from conventional data silos, offering a unified platform where structured and unstructured data coalesce seamlessly. Through distributed computing, parallel processing, and robust data management, Data Cloud sets the stage for a data science revolution. The capabilities and tools that empower data scientists are seamlessly integrated and are designed to handle diverse data types and analytical workloads within Data Cloud. As such, Data Cloud offers various advantages for running data science and ML workloads.

Advantages of Data Cloud for data science

One of the key advantages of Snowflake's Data Cloud is the ability to store and process vast amounts of data without the constraints of hardware limitations. It offers a scalable solution to handling vast volumes of data, enabling data scientists to work with extensive datasets without having to worry about computing or storage constraints. The cloud-based interface provides a collaborative and flexible environment for data scientists and analysts and comes with built-in collaboration features, version control, and support for popular data science libraries and frameworks through Snowpark.

Furthermore, Data Cloud offers a diverse ecosystem of services and resources tailored for data science tasks through managed services that simplify these processes, from data ingestion and preparation to ML model training and deployment. For instance, data pipelines can be automated using serverless computing, and ML models can be trained on powerful GPU instances, leading to faster experimentation and iteration. Data security and compliance are paramount in data science, especially when dealing with sensitive information, and Data Cloud provides different security measures, including encryption, access control, and row-level policies, ensuring that data scientists can work with sensitive data in a secure and compliant manner, adhering to industry regulations and organizational policies. The Snowpark framework is at the center of Snowflake's Data Cloud to support these capabilities. The following section will discuss why Snowpark is used for data science and ML.

Why Snowpark for data science and ML?

Snowpark offers unparalleled integration capabilities for data engineers, enabling seamless interaction with data stored in large volumes and diverse formats. Its versatile API facilitates effortless data exploration, transformation, and manipulation, laying a robust foundation for data science models and ML development and empowering data scientists to harness the full potential of their analytical workflows. Data science teams can now focus on their core tasks without the hassle of infrastructure or environment maintenance.

Snowpark excels in scalability and performance, which is crucial for enterprise data science and ML workloads; leveraging Snowflake's distributed architecture to handle massive datasets and complex computations with remarkable efficiency and the ability to parallelize processing tasks and distribute workloads across multiple nodes ensures lightning-fast execution, even when dealing with petabytes of data. These features, combined with Snowflake's automatic optimization features, allow data scientists to focus on their analyses without being burdened by infrastructure limitations.

Snowpark offers a rich array of advanced analytics capabilities that are indispensable for data science and ML tasks. From statistical analysis to predictive modeling, geospatial analytics, or even data mining, it provides a comprehensive toolkit for data scientists to explore complex patterns and extract valuable insights. Its support for ML libraries and algorithms further amplifies its utility, enabling the development of sophisticated models for classification, regression, and clustering. With the rich features and functionalities mentioned previously, Snowpark provides many benefits for data science and ML workloads. In the next section, we will explore the world of the Snowpark ML library and its different functionalities.

Introduction to Snowpark ML

Snowpark constitutes a compendium of libraries and runtimes within Snowflake, facilitating the secure deployment and processing of non-SQL code by encompassing languages such as Python with the code execution that occurs server-side within the Snowflake infrastructure, all while leveraging a virtual warehouse. The newest addition to the Snowpark libraries is Snowpark ML. Snowpark ML represents a groundbreaking fusion of Snowflake's powerful data processing capabilities and the transformative potential of ML. As the frontier of data science expands, Snowpark ML emerges as a cutting-edge framework that's designed to empower data professionals to harness the full potential of their data within Snowflake's cloud-based environment.

At its core, Snowpark ML is engineered to facilitate seamless integration between Snowflake's data processing capabilities and advanced ML techniques. With Snowpark ML, data scientists, analysts, and engineers can leverage familiar programming languages and libraries to develop sophisticated ML models directly within Snowflake. This integration eliminates the barriers between data storage, processing, and modeling, streamlining the end-to-end data science workflow. Snowpark ML catalyzes innovation, enabling data professionals to efficiently explore, transform, and model data. By bridging the gap between data processing and ML, Snowpark ML empowers organizations to make data-driven decisions, uncover valuable insights, and drive business growth in the digital age. The following figure shows the Snowpark ML framework:

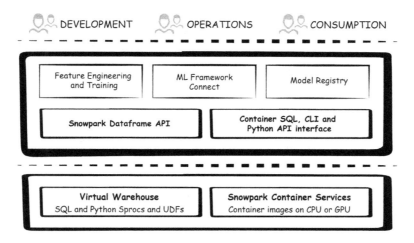

Figure 5.1 – Snowpark ML

The preceding architecture consists of various components that work cohesively together. We will look at each of these components in more detail in the next section.

Snowpark ML API

Similar to Snowpark DataFrame, which helps operate with the data, Snowpark ML provides APIs as a Python library called `snowflake-ml` to support every stage of the ML development and deployment process, allowing support for pre-processing data and training, managing, and deploying ML models all within Snowflake:

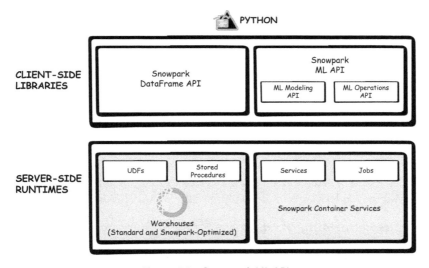

Figure 5.2 – Snowpark ML API

The Snowpark ML API consists of Snowpark ML modeling for developing and training the models and Snowpark ML Ops for monitoring and operating the model. The `snowflake.ml.modeling` module provides APIs for pre-processing, feature engineering, and model training based on familiar libraries, such as scikit-learn and XGBoost. The complete end-to-end ML experience can be done using Snowpark. We'll cover this in the next section.

End-to-end ML with Snowpark

The quest for seamless, end-to-end ML solutions has become paramount, and Snowpark offers a comprehensive ecosystem for end-to-end ML. This section delves into the intricate world of leveraging Snowpark to craft end-to-end ML pipelines, from data ingestion and preprocessing to model development, training, and deployment, unveiling the seamless process of developing ML within Snowflake's robust framework.

ML processes involve a systematic approach to solving complex problems through data processing. This typically includes stages such as defining the problem, collecting and preparing data, **exploratory data analysis** (EDA), feature engineering, model selection, training, evaluation, and deployment, with each operation being crucial and iterative. It allows data scientists to refine their approaches based on insights gained along the way. The process is often cyclical, with continuous iterations to improve models and predictions:

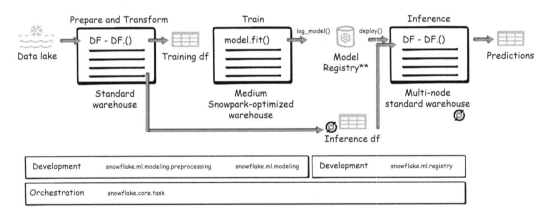

Figure 5.3 – End-to-end ML flow

We can broadly classify the ML stages as preparing and transforming, training and building the model, and interfering with the model to obtain prediction results. We will discuss each of these steps next.

Preparing and transforming the data

Raw data is often messy, containing missing values, outliers, and inconsistencies. Getting the correct data from multiple systems usually consumes most of the data scientist's time. Snowflake solves this problem by providing a governed Data Cloud paradigm that supports all types of data and provides a unified place to instantly store and consume relevant data to unlock ML models' power. The data preparation and transformation process involves EDA, cleaning, and processing, and ends with feature engineering. This step also consists of data engineering pipelines, which help apply data transformations to prepare the data for the next step. For data pre-processing, `snowflake.ml.modeling`, preprocessing, and Snowpark functions can be used to transform the data.

EDA

EDA is a critical step that involves preliminary investigations to understand the data's structure, patterns, and trends as it helps uncover hidden patterns and guide feature selection. Data scientists and analysts collaborate closely with business stakeholders to define the specific questions that need to be answered or the problems that need to be solved, which guides them in selecting the relevant data. Through charts, graphs, and statistical summaries, data scientists can identify patterns, trends, correlations, and outliers within the dataset, all of which provide valuable insights into the data's distribution and help them understand the data better to build feature selection.

Data cleaning and preprocessing

Data cleaning involves handling missing data, correcting errors, and ensuring consistency. The data is suitable for training the model through preprocessing techniques such as normalization, scaling, and transformations, along with various sampling techniques that are applied to evaluate a subset of the data, providing insights into its richness and variability.

Feature engineering

Feature engineering involves creating new features or modifying existing ones to enhance the performance of ML models. It requires domain expertise to identify relevant features that can improve predictive accuracy. Performing feature engineering on the centralized data in Snowflake accelerates model development, reduces costs, and enables the reuse of new features. Some techniques, such as creating interaction terms and transforming variables, extract meaningful information from raw data, making it more informative for modeling.

Training and building the model

Once the data is ready and the features have been built, the next step is to train and develop the model. In this stage, the data scientist trains various models, such as regression, classification, clustering, or deep learning, depending on the nature of the problem, by passing a subset of the data, or training set, through the modeling function to derive a predictive function. The model is developed using statistical methods for hypothesis testing and inferential statistics. Advanced techniques, such as ensemble methods, neural networks, and natural language processing, are also used, depending on the data.

Once the model has been developed, it's tested on data that wasn't part of the training set to determine its effectiveness, which is usually measured in terms of its predictive strength and robustness, and the model is optimized with hyperparameter tuning. Cross-validation techniques optimize the model's performance, ensuring accurate predictions and valuable insights.

This combination of steps enables data scientists to conduct in-depth feature engineering, tune hyperparameters, and iteratively create and assess ML models. Intuitions become accurate predictions as data scientists experiment with various algorithms, evaluating the performance of each model and adjusting parameters on their chosen model to optimize the code for their specific datasets. `snowflake.ml.modeling` can be used for training by utilizing the `fit()` method for an algorithm such as XGBoost.

Inference

Once the models have been trained, Snowpark ML supports their seamless deployment and inference. Models can be deployed for inference, enabling organizations to make data-driven decisions based on predictive insights. Snowpark ML has a model registry to manage and organize Snowpark models throughout their life cycle. The model registry supports versioning of the models and stores metadata information about the models, hyperparameters, and evaluation metrics, facilitating experimentation and model comparison. It also supports model monitoring and auditing and aids in collaboration between data scientists working on the model. The model registry is part of Snowpark MLOps and can be accessed through `snowpark.ml.registry`. The pipelines can be orchestrated using Snowflake Tasks.

Now that we have established the foundations of Snowpark ML, its place in ML, and how Snowpark supports data science workloads, we will dive deep into the complete data science scenario with Snowpark. The following section will focus on exploring and preparing the data.

> **A note on data engineering**
>
> In the next section, we'll conduct exploration, transformation, and feature engineering using Snowpark Python and pandas. As we proceed to build models with SnowparkML, we will incorporate some of the steps discussed earlier in this section.

Exploring and preparing data

In the first step of the ML process, we must explore and prepare the data in Snowflake using Snowpark to make it available for training the ML models. We will work with the Bike Sharing dataset from Kaggle, which offers an hourly record of rental data for 2 years. The primary objective is to forecast the number of bikes rented each hour for a specific timeframe based solely on the information available before the rental period. In essence, the model will harness the power of historical data to predict future bike rental patterns using Snowpark. More information about the particular dataset has been provided in the respective GitHub repository (`https://github.com/PacktPublishing/The-Ultimate-Guide-To-Snowpark`).

Data exploration allows us to dissect the data to uncover intricate details that might otherwise stay hidden, acting as the foundation for our entire analysis. We will start the process by loading the dataset into a Snowpark DataFrame:

```
df_table=session.table("BSD_TRAINING")
```

Once the data has been successfully loaded, the subsequent imperative is to gain a comprehensive understanding of the dataset's scale:

```
number_of_rows = df_table.count()
number_of_columns = len(df_table.columns)
```

Fortunately, Snowpark provides functions specifically designed to facilitate this critical task:

```
Total No of rows in dataset :  10886
Total No of columns in dataset :  12
```

Figure 5.4 – Total number of columns

Now that we know the scale of the data, let's get a sense of it by looking at a few rows of the dataset:

```
df_table.sample(n=2).show()
```

This returns the two rows from the data for analysis:

```
---------------------------------------------------------------------------------------------------------------------------------------------------------
|"SEASON"  |"HOLIDAY"  |"WORKINGDAY"  |"WEATHER"  |"TEMP"  |"ATEMP"  |"HUMIDITY"  |"WINDSPEED"  |"CASUAL"  |"REGISTERED"  |"COUNT"  |"DATETIME"           |
---------------------------------------------------------------------------------------------------------------------------------------------------------
|3         |0          |1             |3          |24.6    |27.275   |88          |19.9995      |19        |122           |141      |2012-09-18 16:00:00  |
|2         |0          |0             |2          |14.76   |17.425   |81          |15.0013      |7         |36            |43       |2011-04-16 00:00:00  |
---------------------------------------------------------------------------------------------------------------------------------------------------------
```

Figure 5.5 – Two rows of data

As depicted in the preceding figure, the COUNT column is a straightforward aggregation of CASUAL and REGISTERED. In data science, these types of variables are commonly referred to as "leakage variables." When we construct our models, we'll delve deeper into strategies for managing these variables. Date columns consistently present an intriguing and complex category to grapple with. Within this dataset, there is potential to create valuable new features derived from the DATETIME column, which could significantly influence our response variables. Before we start with data cleansing and the feature engineering process, let's see the column type to understand and make more informed decisions:

```
import pprint
data_types = df_table.schema
data_types = df_table.schema.fields
pprint.pprint(data_types)
```

This will give us the schema and the data types for each field so that we can understand the data better:

```
[StructField('SEASON', LongType(), nullable=True),
 StructField('HOLIDAY', LongType(), nullable=True),
 StructField('WORKINGDAY', LongType(), nullable=True),
 StructField('WEATHER', LongType(), nullable=True),
 StructField('TEMP', DoubleType(), nullable=True),
 StructField('ATEMP', DoubleType(), nullable=True),
 StructField('HUMIDITY', LongType(), nullable=True),
 StructField('WINDSPEED', DoubleType(), nullable=True),
 StructField('CASUAL', LongType(), nullable=True),
 StructField('REGISTERED', LongType(), nullable=True),
 StructField('COUNT', LongType(), nullable=True),
 StructField('DATETIME', TimestampType(tz=ntz), nullable=True)]
```

Figure 5.6 – Schema information

Now that we are equipped with basic information about the data, let's start finding the missing values in the data.

Missing value analysis

Addressing missing values is a fundamental preprocessing step in ML. Incomplete data can disrupt model training and hinder predictive accuracy, potentially leading to erroneous conclusions or suboptimal performance. By systematically imputing or filling these gaps, we can bolster the integrity of our dataset, providing ML algorithms with a more comprehensive and coherent dataset for more robust and reliable analyses and predictions. This practice is akin to affording our models the necessary information to make sound, data-driven decisions. Let's check for any missing values in our dataset:

```
from snowflake.snowpark.functions import count, col
data_types = df_table.schema
print(data_types)
for column in df_table.columns:
    print(f"Null values in {column} is {number_of_rows - df_table.
agg(count(col(column))).collect()[0][0]}")
```

The preceding code helps us find out whether any values are empty or missing:

```
StructType([StructField('SEASON', LongType(), nullable=True), StructField('HOLIDAY', LongType(), nullable=True),
Null values in SEASON is 0
Null values in HOLIDAY is 0
Null values in WORKINGDAY is 0
Null values in WEATHER is 0
Null values in TEMP is 0
Null values in ATEMP is 0
Null values in HUMIDITY is 0
Null values in WINDSPEED is 0
Null values in CASUAL is 0
Null values in REGISTERED is 0
Null values in COUNT is 0
Null values in DATETIME is 0
```

Figure 5.7 – Missing value analysis

Our initial examination of missing values in the column shows no missing values in our dataset. However, a closer examination reveals the presence of numerous 0s within the WINDSPEED column, which is indicative of potentially missing values. Logically, windspeed cannot equate to zero, implying that each 0 within the column signifies a missing value:

```
print(f"Zero Values in windspeed column is {df_table.filter(df_
table['WINDSPEED']==0).count()}")
```

This will print out the following output:

```
Zero Values in windspeed column is 1313
```

Figure 5.8 – Output value

We can see that there are 1313 values in the WINDSPEED column. With this column harboring missing data, the subsequent challenge is determining an effective strategy for imputing these missing values. As is widely acknowledged, various methods exist for addressing missing data within a column. In our case, we'll employ a straightforward imputation, substituting the 0s with the mean value of the column:

```
from snowflake.snowpark.functions import iff, avg
wind_speed_mean = df_train.select(mean("windspeed")).collect()[0][0]
df_train = df_train.replace({0:wind_speed_mean}, subset=["windspeed"])
df_train.show()
df_train.write.mode("overwrite").save_as_table("model_data")
```

The preceding code replaces the 0s with the mean value of the column:

"SEASON"	"HOLIDAY"	"WORKINGDAY"	"WEATHER"	"TEMP"	"ATEMP"	"HUMIDITY"	"WINDSPEED"	"CASUAL"	"REGISTERED"	"COUNT"
1	0	0	1	9.84	14.395	81	12.7993954069447	3	13	16
1	0	0	1	9.02	13.635	80	12.7993954069447	8	32	40
1	0	0	1	9.02	13.635	80	12.7993954069447	5	27	32
1	0	0	1	9.84	14.395	75	12.7993954069447	3	10	13
1	0	0	1	9.84	14.395	75	12.7993954069447	0	1	1
1	0	0	2	9.84	12.88	75	6.0032	0	1	1
1	0	0	1	9.02	13.635	80	12.7993954069447	2	0	2
1	0	0	1	8.2	12.88	86	12.7993954069447	1	2	3
1	0	0	1	9.84	14.395	75	12.7993954069447	1	7	8
1	0	0	1	13.12	17.425	76	12.7993954069447	8	6	14

Figure 5.9 – Pre-processed data

This concludes our preprocessing journey. Next, we'll perform outlier analysis.

Outlier analysis

The process of detecting and removing outliers is pivotal in enhancing model accuracy and robustness. Outliers are data points that significantly deviate from most datasets, often stemming from errors, anomalies, or rare events. These aberrations can unduly influence model training, leading to skewed predictions or reduced generalization capabilities. By identifying and eliminating outliers, we can improve the quality and reliability of our models and ensure that they are better equipped to discern meaningful patterns within the data. This practice fosters more accurate predictions and a higher level of resilience, ultimately contributing to the overall success of ML endeavors.

We will be transforming the DataFrame into a pandas DataFrame so that we can conduct insightful analyses, including constructing visualizations to extract meaningful patterns. Our initial focus is on the COUNT column as the response variable. Before model development, it is imperative to ascertain whether the COUNT column contains any outlier values:

```
import seaborn as sns
import matplotlib.pyplot as plt
f, axes = plt.subplots(1, 2)
sns.boxplot(x=df_table.to_pandas()['COUNT'], ax=axes[0])
sns.boxplot(x=df_without_outlier.to_pandas()['COUNT'], ax=axes[1])
plt.show()
```

The preceding code generates a plot using the seaborn and the matplotlib library to help us find the outliers:

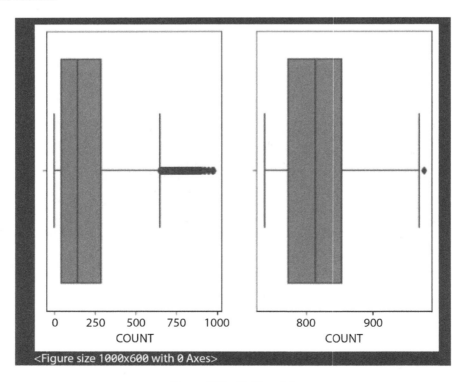

Figure 5.10 – Outlier plot

As we can see, the `COUNT` column exhibits outlier data points that can potentially negatively impact model performance if they're not adequately addressed. Mitigating outliers is a critical preprocessing step. One widely adopted approach involves removing data points that lie beyond a predefined threshold or permissible range, as outlined here:

```
from snowflake.snowpark.functions \
    import mean, stddev, abs, date_part

mean_value = df_table.select(mean("count")).collect()[0][0]
print(mean_value)
std_value = df_table.select(stddev("count")).collect()[0][0]
print(std_value)
df_without_outlier = df_table.filter(
    (abs(df_table["count"] - mean_value)) >= (3 * std_value))
df_without_outlier.show()
```

The preceding code uses the Snowpark library to analyze a dataset stored in `df_table`. It calculates the mean (average) and standard deviation (a measure of data spread) of the `'count'` column in the dataset. Then, it identifies and removes outliers from the dataset. Outliers are data points that significantly differ from the average. In this case, it defines outliers as data points more than three times the standard deviation away from the mean. After identifying these outliers, it displays the dataset without the outlier values, using `df_without_outlier.show()` to help with further analysis:

191.574132
181.14445382898148

"SEASON"	"HOLIDAY"	"WORKINGDAY"	"WEATHER"	"TEMP"	"ATEMP"	"HUMIDITY"	"WINDSPEED"	"CASUAL"	"REGISTERED"	"COUNT"	"DATETIME"	"HOUR"	"MONTH"	"DATE"	"WEEKDAY"
1	0	1	1	28.7	31.82	28	6.0032	140	642	782	2012-03-14 17:00:00	17	3	2012-03-14	3
1	0	1	1	28.7	31.82	32	0.0	102	647	749	2012-03-14 18:00:00	18	3	2012-03-14	3
1	0	1	1	27.06	31.06	44	19.0012	98	648	746	2012-03-15 18:00:00	18	3	2012-03-15	4
1	0	1	1	26.24	31.06	57	16.9979	120	681	801	2012-03-19 18:00:00	18	3	2012-03-19	1
2	0	1	1	25.42	30.305	17	12.998	123	634	757	2012-04-03 17:00:00	17	4	2012-04-03	2
2	0	1	1	24.6	30.305	26	11.0014	139	661	800	2012-04-03 18:00:00	18	4	2012-04-03	2
2	0	1	1	27.06	31.06	41	19.9995	99	645	744	2012-04-04 17:00:00	17	4	2012-04-04	3
2	0	1	1	27.06	31.06	41	16.9979	113	646	759	2012-04-04 18:00:00	18	4	2012-04-04	3
2	0	1	1	20.5	24.24	31	11.0014	145	677	822	2012-04-05 17:00:00	17	4	2012-04-05	4
2	0	1	1	22.96	26.515	21	35.0008	88	656	744	2012-04-10 17:00:00	17	4	2012-04-10	2

Figure 5.11 – Outliers removed

Now that we have taken care of the outliers, we can perform correlation analysis.

Correlation analysis

Identifying correlations among variables is of paramount importance for several vital reasons. Correlations provide valuable insights into how different features in the dataset relate to each other. By understanding these relationships, ML models can make more informed predictions as they leverage the strength and direction of correlations to uncover patterns and dependencies. Moreover, identifying and quantifying correlations aids feature selection, where irrelevant or highly correlated features can be excluded to enhance model efficiency and interpretability. It also helps identify potential multicollinearity issues, where two or more features are highly correlated, leading to unstable model coefficients. Recognizing and harnessing correlations empowers ML models to make better predictions and yield more robust results, making it a fundamental aspect of modeling.

Since we've already transformed our Snowpark DataFrame into a pandas DataFrame, we can readily create a correlation matrix, a fundamental tool for exploring relationships between variables. Only the snippet to generate a correlation matrix is demonstrated here but the complete code is available in `Chapter_5.ipynb`:

```
corr_matrix = df_without_outlier.to_pandas().corr()
plt.figure(figsize=(12, 6))
sns.heatmap(corr_matrix, cmap='coolwarm', annot=True)
```

The preceding code generates the correlation matrix as a heatmap visualization:

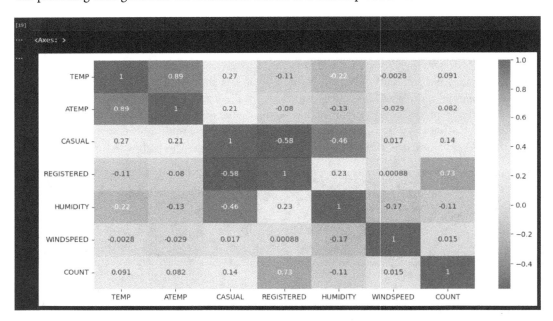

Figure 5.12 – Correlation matrix heatmap

This heatmap visualization reveals a substantial correlation between the TEMP and ATEMP variables, signifying a condition known as multicollinearity. Multicollinearity occurs when two or more predictors in a model are highly correlated, distorting the model's interpretability and stability. To mitigate this issue and ensure the reliability of our analysis, we have opted to retain the TEMP variable while removing ATEMP from consideration in our subsequent modeling endeavors. This strategic decision is made to maintain model robustness and effectively capture the essence of the data without the confounding effects of multicollinearity.

Leakage variables

Leakage variables in data science inadvertently include information that would not be available during prediction or decision-making in a real-world scenario. Eliminating them is crucial because using leakage variables can lead to overly optimistic model performance and unreliable results. It's essential to detect and exclude these variables during data preprocessing to ensure that our models make predictions based on the same information that would be accessible. By doing so, we prevent the risk of building models that work well on historical data but fail to perform in real-world situations, which is a crucial goal in data science.

As mentioned previously, the CASUAL, REGISTERED, and COUNT columns exhibit high collinearity, with COUNT being an explicit summation of CASUAL and REGISTERED. This redundancy renders the inclusion of all three variables undesirable, resulting in a leakage variable situation. To preserve the integrity of our model-building process, we shall eliminate CASUAL and REGISTERED from our feature set, thereby mitigating any potential confounding effects and ensuring the model's ability to make predictions based on the most relevant and non-redundant information. The next step is to perform feature engineering with the prepared data.

Feature engineering

Feature engineering in ML is like crafting the perfect tool for a specific job. It involves selecting, transforming, or creating new features (variables) from the available data to make it more suitable for ML algorithms. This process is crucial because it helps the models better understand the patterns and relationships in the data, leading to improved predictions and insights. By carefully engineering features, we can uncover hidden information, reduce noise, and enhance the model's performance, making it a vital step in building effective and accurate ML systems.

Analyzing the data shows that the DATETIME column is a promising candidate for feature engineering within this dataset. Given the dependency of the predictive outcome on temporal factors such as the time of day and day of the week, deriving time-related features assumes paramount significance. Extracting these temporal features is pivotal as it enhances the model's performance and elevates the overall predictive accuracy by capturing essential nuances about the dataset's material characteristics:

```
from snowflake.snowpark.functions import hour, month,to_date,dayofweek
df_table = df_table.with_column("hour", hour("DATETIME"))
df_table = df_table.with_column("month", month("DATETIME"))
df_table = df_table.with_column("date", to_date("DATETIME"))
df_table = df_table.with_column("weekday", dayofweek("DATETIME"))
df_table.show()
```

The preceding code enriches a DataFrame by creating new columns that capture specific time and date details from the DATETIME column:

"WEATHER"	"TEMP"	"ATEMP"	"HUMIDITY"	"WINDSPEED"	"CASUAL"	"REGISTERED"	"COUNT"	"DATETIME"	"HOUR"	"MONTH"	"DATE"	"WEEKDAY"
1	9.84	14.395	81	0.0	3	13	16	2011-01-01 00:00:00	0	1	2011-01-01	6
1	9.02	13.635	80	0.0	8	32	40	2011-01-01 01:00:00	1	1	2011-01-01	6
1	9.02	13.635	80	0.0	5	27	32	2011-01-01 02:00:00	2	1	2011-01-01	6
1	9.84	14.395	75	0.0	3	10	13	2011-01-01 03:00:00	3	1	2011-01-01	6
1	9.84	14.395	75	0.0	0	1	1	2011-01-01 04:00:00	4	1	2011-01-01	6
2	9.84	12.88	75	6.0032	0	1	1	2011-01-01 05:00:00	5	1	2011-01-01	6
1	9.02	13.635	80	0.0	2	0	2	2011-01-01 06:00:00	6	1	2011-01-01	6
1	8.2	12.88	86	0.0	1	2	3	2011-01-01 07:00:00	7	1	2011-01-01	6
1	9.84	14.395	75	0.0	1	7	8	2011-01-01 08:00:00	8	1	2011-01-01	6
1	13.12	17.425	76	0.0	8	6	14	2011-01-01 09:00:00	9	1	2011-01-01	6

Figure 5.13 – DATETIME data

The hour column tells us the hour of the day, the month column identifies the month, the date column extracts the date itself, and the weekday column signifies the day of the week. These additional columns provide a more comprehensive view of the time-related information within the dataset, enhancing its potential for in-depth analysis and ML applications.

This step concludes our data preparation and exploration journey. The following section will use this prepared data to build and train our model using Snowpark.

> **A note on the model-building process**
>
> In our model-building process, we won't be incorporating all the steps we've discussed thus far. Instead, we'll focus on two significant transformations to showcase Snowpark ML pipelines. Additionally, the accompanying notebook (chapter_5.ipynb) illustrates model building using Python's scikit-learn library and how to call them as stored procedures. This will allow you to compare and contrast how the model-building process is simplified through Snowpark ML. To follow through the chapter, you can skip the model building process using the scikit-learn section and directly go to the Snowpark ML section in the notebook.

Training ML models in Snowpark

Now that we have prepared our dataset, the pinnacle of our journey involves the model-building process, for which we will be leveraging the power of Snowpark ML. Snowpark ML emerges as a recent addition to the Snowpark arsenal, strategically deployed to streamline the intricacies of the model-building process. Its elegance becomes apparent when we engage in a comparative exploration of the model-building procedure through the novel ML library. We will start by developing the pipeline that we'll use to train the model using the data we prepared previously:

```
import snowflake.ml.modeling.preprocessing as snowml
from snowflake.ml.modeling.pipeline import Pipeline
import joblib
df = session.table("BSD_TRAINING")
df = df.drop("DATETIME","DATE")
CATEGORICAL_COLUMNS = ["SEASON","WEATHER"]
```

```
CATEGORICAL_COLUMNS_OHE = ["SEASON_OE","WEATHER_OE"]
MIN_MAX_COLUMNS = ["TEMP"]
import numpy as np
categories = {
    "SEASON": np.array([1,2,3,4]),
    "WEATHER": np.array([1,2,3,4]),
}
preprocessing_pipeline = Pipeline(
    steps=[
        (
            "OE",
            snowml.OrdinalEncoder(
                input_cols=CATEGORICAL_COLUMNS,
                output_cols=CATEGORICAL_COLUMNS_OHE,
                categories=categories
            )
        ),
        (
            "MMS",
            snowml.MinMaxScaler(
                clip=True,
                input_cols=MIN_MAX_COLUMNS,
                output_cols=MIN_MAX_COLUMNS,
            )
        )
    ]
)
PIPELINE_FILE = 'preprocessing_pipeline.joblib'
joblib.dump(preprocessing_pipeline, PIPELINE_FILE)
transformed_df = preprocessing_pipeline.fit(df).transform(df)
transformed_df.show()
session.file.put(PIPELINE_FILE,"@snowpark_test_stage",overwrite=True)
```

The preceding code creates a preprocessing pipeline for the dataset by using various Snowpark ML functions. The preprocessing and pipeline modules are imported as these are essential for developing and training the model:

"TEMP"	"SEASON_OE"	"WEATHER_OE"	"SEASON"	"HOLIDAY"	"WORKINGDAY"	"WEATHER"	"ATEMP"	"HUMIDITY"	"WINDSPEED"	"CASUAL"	"REGISTERED"	"COUNT"
0.2244897959183673	0.0	0.0	1	0	0	1	14.395	81	0.0	3	13	16
0.20408163265306123	0.0	0.0	1	0	0	1	13.635	80	0.0	8	32	40
0.20408163265306123	0.0	0.0	1	0	0	1	13.635	80	0.0	5	27	32
0.2244897959183673	0.0	0.0	1	0	0	1	14.395	75	0.0	3	10	13
0.2244897959183673	0.0	0.0	1	0	0	1	14.395	75	0.0	0	1	1
0.2244897959183673	0.0	1.0	1	0	0	2	12.88	75	6.0032	0	1	1
0.20408163265306123	0.0	0.0	1	0	0	1	13.635	80	0.0	2	0	2
0.18367346938775508	0.0	0.0	1	0	0	1	12.88	86	0.0	1	2	3
0.2244897959183673	0.0	0.0	1	0	0	1	14.395	75	0.0	1	7	8
0.3061224489795918	0.0	0.0	1	0	0	1	17.425	76	0.0	8	6	14

Figure 5.14 – Transformed data

The pipeline includes ordinal encoding for categorical columns (SEASON and WEATHER) and min-max scaling for numerical columns (TEMP). The pipeline is saved into the stage using the joblib library, which can be utilized for consistent preprocessing in future analyses. Now that we have the pipeline code ready, we will build the features that are required for the model:

```
CATEGORICAL_COLUMNS = ["SEASON","WEATHER"]
CATEGORICAL_COLUMNS_OHE = ["SEASON_OE","WEATHER_OE"]
MIN_MAX_COLUMNS = ["TEMP","ATEMP"]
FEATURE_LIST = \
    ["HOLIDAY","WORKINGDAY","HUMIDITY","TEMP","ATEMP","WINDSPEED"]

LABEL_COLUMNS = ['COUNT']
OUTPUT_COLUMNS = ['PREDICTED_COUNT']

PIPELINE_FILE = 'preprocessing_pipeline.joblib'
preprocessing_pipeline = joblib.load(PIPELINE_FILE)
```

The preceding code defines lists representing categorical columns, one-hot encoded categorical columns, and columns for min-max scaling. It also specifies a feature list, label columns, and output columns for an ML model. The preprocessing_pipeline.joblib file is loaded and assumed to contain a previously saved preprocessing pipeline. These elements collectively prepare the necessary data and configurations for subsequent ML tasks, ensuring consistent handling of categorical variables, feature scaling, and model predictions based on the pre-established pipeline. We will now split the data into training and testing sets:

```
bsd_train_df, bsd_test_df = df.random_split(
    weights=[0.7,0.3], seed=0)
train_df = preprocessing_pipeline.fit(
    bsd_train_df).transform(bsd_train_df)
test_df = preprocessing_pipeline.transform(bsd_test_df)
train_df.show()
test_df.show()
```

The preceding code divides the dataset into training (70%) and testing (30%) sets using a random split. It applies the previously defined preprocessing pipeline to transform both sets, displaying the transformed training and testing datasets and ensuring consistent preprocessing for model training and evaluation. The output shows the different training and testing data:

"TEMP"	"SEASON_OE"	"WEATHER_OE"	"SEASON"	"HOLIDAY"	"WORKINGDAY"	"WEATHER"	"ATEMP"	"HUMIDITY"	"WINDSPEED"	"CASUAL"	"REGISTERED"	"COUNT"
0.2340425531914894	0.0	0.0	1	0	0	1	14.395	81	0.0	3	13	16
0.21276595744680848	0.0	0.0	1	0	0	1	13.635	80	0.0	8	32	40
0.21276595744680848	0.0	0.0	1	0	0	1	13.635	80	0.0	5	27	32
0.2340425531914894	0.0	0.0	1	0	0	1	14.395	75	0.0	3	10	13
0.2340425531914894	0.0	1.0	1	0	0	2	12.88	75	6.0032	0	1	1
0.21276595744680848	0.0	0.0	1	0	0	1	13.635	80	0.0	2	0	2
0.2340425531914894	0.0	0.0	1	0	0	1	14.395	75	0.0	1	7	8
0.3191489361702128	0.0	0.0	1	0	0	1	17.425	76	0.0	8	6	14
0.3617021276595745	0.0	0.0	1	0	0	1	16.665	81	19.0012	26	30	56
0.425531914893617	0.0	0.0	1	0	0	1	21.21	77	19.0012	29	55	84

"TEMP"	"SEASON_OE"	"WEATHER_OE"	"SEASON"	"HOLIDAY"	"WORKINGDAY"	"WEATHER"	"ATEMP"	"HUMIDITY"	"WINDSPEED"	"CASUAL"	"REGISTERED"	"COUNT"
0.2340425531914894	0.0	0.0	1	0	0	1	14.395	75	0.0	0	1	1
0.19148936170212763	0.0	0.0	1	0	0	1	12.88	86	0.0	1	2	3
0.3829787234042553	0.0	0.0	1	0	0	1	19.695	76	16.9979	12	24	36
0.46808510638297873	0.0	1.0	1	0	0	2	22.725	72	19.9995	47	47	94
0.46808510638297873	0.0	1.0	1	0	0	2	22.725	72	19.0012	35	71	106
0.425531914893617	0.0	1.0	1	0	0	2	21.21	82	19.9995	41	52	93
0.425531914893617	0.0	2.0	1	0	0	3	21.21	88	16.9979	6	31	37
...												
0.46808510638297873	0.0	1.0	1	0	0	2	22.725	88	19.9995	15	24	39
0.44680851063829785	0.0	1.0	1	0	0	2	21.97	94	16.9979	1	16	17

Figure 5.15 – Training and testing dataset

Next, we'll train the model with the training data:

```
from snowflake.ml.modeling.linear_model import LinearRegression

regressor = LinearRegression(
    input_cols=CATEGORICAL_COLUMNS_OHE+FEATURE_LIST,
    label_cols=LABEL_COLUMNS,
    output_cols=OUTPUT_COLUMNS
)
# Train
regressor.fit(train_df)
# Predict
result = regressor.predict(test_df)
result.show()
```

The LinearRegression class defines the model, specifying the input columns (categorical columns after one-hot encoding and additional features), label columns (the target variable – that is, COUNT), and output columns for predictions. The model is trained on the transformed training dataset using fit, and then predictions are generated for the transformed testing dataset using predict. The resulting predictions are displayed, assessing the model's performance on the test data:

"WINDSPEED"	"TEMP"	"CASUAL"	"ATEMP"	"HOLIDAY"	"COUNT"	"WORKINGDAY"	"HUMIDITY"	"WEATHER_OE"	"REGISTERED"	"WEATHER"	"SEASON"	"SEASON_OE"	"PREDICTED_COUNT"
0.0	0.2340425531914894	0	14.395	0	1	0	75	0.0	1	1	1	0.0	33.90384901378961
0.0	0.19148936170212763	1	12.88	0	3	0	86	0.0	2	1	1	0.0	-10.36060733142813
16.9979	0.3829787234042553	12	19.695	0	36	0	76	0.0	24	1	1	0.0	86.33327234926997
19.9995	0.46808510638297873	47	22.725	0	94	0	72	1.0	47	2	1	0.0	131.80551630295983
19.0012	0.46808510638297873	35	22.725	0	106	0	72	1.0	71	2	1	0.0	130.89217838200688
19.9995	0.425531914893617	41	21.21	0	93	0	82	1.0	52	2	1	0.0	90.53603978393944
16.9979	0.425531914893617	6	21.21	0	37	0	88	2.0	31	3	1	0.0	78.10684082594925
15.0013	0.40425531914893614	11	20.455	0	28	0	94	1.0	17	2	1	0.0	44.39336148403612
19.9995	0.46808510638297873	15	22.725	0	39	0	88	1.0	24	2	1	0.0	83.91784036378053
16.9979	0.44680851063829785	1	21.97	0	17	0	94	1.0	16	2	1	0.0	57.55973646297843

Figure 5.16 – Predicted output

The next step is to calculate various performance metrics to evaluate the accuracy of the linear regression model's predictions:

```
from snowflake.ml.modeling.metrics import mean_squared_error,
explained_variance_score, mean_absolute_error, mean_absolute_
percentage_error, d2_absolute_error_score, d2_pinball_score

mse = mean_squared_error(df=result,
    y_true_col_names="COUNT",
    y_pred_col_names="PREDICTED_COUNT")
evs = explained_variance_score(df=result,
    y_true_col_names="COUNT",
    y_pred_col_names="PREDICTED_COUNT")
mae = mean_absolute_error(df=result,
    y_true_col_names="COUNT",
    y_pred_col_names="PREDICTED_COUNT")
mape = mean_absolute_percentage_error(df=result,
    y_true_col_names="COUNT",
    y_pred_col_names="PREDICTED_COUNT")
d2aes = d2_absolute_error_score(df=result,
    y_true_col_names="COUNT",
    y_pred_col_names="PREDICTED_COUNT")
d2ps = d2_pinball_score(df=result,
    y_true_col_names="COUNT",
    y_pred_col_names="PREDICTED_COUNT")

print(f"Mean squared error: {mse}")
print(f"explained_variance_score: {evs}")
print(f"mean_absolute_error: {mae}")
print(f"mean_absolute_percentage_error: {mape}")
print(f"d2_absolute_error_score: {d2aes}")
print(f"d2_pinball_score: {d2ps}")
```

The preceding code calculates various performance metrics to assess the accuracy of the linear regression model's predictions. Metrics such as mean squared error, explained variance score, mean absolute error, mean fundamental percentage error, d2 definitive error score, and d2 pinball score are computed based on the actual (COUNT) and predicted (PREDICTED_COUNT) values stored in the result DataFrame:

```
Mean squared error: 19391.524207020506
explained_variance_score: 0.40498512484914284
mean_absolute_error: 103.88982886764924
mean_absolute_percentage_error: 3.4279455496659987
d2_absolute_error_score: 0.24728138773207753
d2_pinball_score: 0.24728138773207753
```

Figure 5.17 – Performance metrics

These performance metrics provide a comprehensive evaluation of the model's performance across different aspects of prediction accuracy.

> **Model results and efficiency**
>
> The presented model metrics might need to showcase more exceptional results. It's crucial to emphasize that the primary objective of this case study is to elucidate the model-building process and highlight the facilitative role of Snowpark ML. The focus of this chapter has been on illustrating the construction of a linear regression model.

The efficiency of Snowpark ML

In delving into the intricacies of the model-building process facilitated by Snowpark ML, the initial standout feature is its well-thought-out design. A notable departure from the conventional approach is evident as Snowpark ML closely mirrors the streamlined methodology found in scikit-learn. A significant advantage is eliminating the need to create separate **user-defined functions** (UDFs) and stored procedures, streamlining the entire model-building workflow.

It's crucial to recognize that Snowpark ML seamlessly integrates with scikit-learn while adhering to similar conventions in the model construction process. A noteworthy distinction is a prerequisite in scikit-learn for data to be passed as a pandas DataFrame. Consequently, the Snowflake table must be converted into a pandas DataFrame before you can initiate the model-building phase. However, it's imperative to be mindful of potential memory constraints, especially when dealing with substantial datasets. Converting a large table into a pandas DataFrame demands a significant amount of memory since the entire dataset is loaded into memory.

In contrast, Snowpark ML provides a more native and memory-efficient approach to the model-building process. This native integration with Snowflake's environment not only enhances the efficiency of the workflow but also mitigates memory-related challenges associated with large datasets. The utilization of Snowpark ML emerges as a strategic and seamless choice for executing complex model-building tasks within the Snowflake ecosystem.

Summary

Snowpark ML emerges as a versatile and powerful tool for data scientists, enabling them to tackle complex data science tasks within Snowflake's unified data platform. Its integration with popular programming languages, scalability, and real-time processing capabilities make it invaluable for various applications, from predictive modeling to real-time analytics and advanced AI tasks. With Snowpark ML, organizations can harness the full potential of their data, drive innovation, and gain a competitive edge in today's data-driven landscape.

In the next chapter, we will continue by deploying the model in Snowflake and operationalizing it.

Deploying and Managing ML Models with Snowpark

The seamless deployment and effective management of models have become pivotal components of developing data science with Snowpark. The previous chapter covered how to prepare the data and train the model. This chapter delves into the intricacies of leveraging Snowpark to deploy and manage **machine learning** (**ML**) models efficiently, from deployment to integration with feature stores and model registries, exploring the essentials of streamlining ML models in Snowpark.

In this chapter, we're going to cover the following main topics:

- Deploying ML models in Snowpark
- Managing Snowpark model data

Technical requirements

Please refer to the *Technical requirements* section in the previous chapter for environment setup.

Supporting materials are available at `https://github.com/PacktPublishing/ The-Ultimate-Guide-To-Snowpark`.

Deploying ML models in Snowpark

In the preceding chapter, we learned about how to develop ML models. Now that the models are ready, we must deploy them into Snowpark. To make it easier for developers to deploy the models, the Snowpark ML library consists of functions that encompass the introduction of a new development interface and additional functionalities aimed at securely facilitating the deployment of both features and models. Snowpark MLOps seamlessly complements the Snowpark ML Development API by offering advanced model management capabilities and integrated deployment functionalities within the Snowflake ecosystem. In the following subsections, we will explore the model registry and deploy the model for inference to obtain predictions.

Snowpark ML model registry

A **model registry** is a centralized repository that enables model developers to organize, share, and publish ML models efficiently. It streamlines collaboration among teams and stakeholders, facilitating the collaborative management of the lifecycle of all models within an organization. Organizing models is crucial for tracking various versions, quickly identifying the latest, and gaining insights into each model's hyperparameters. A well-structured model registry enhances reproducibility and compelling comparison of results. It also allows tracking and analyzing model accuracy metrics, empowering informed decisions and continuous improvement. The following diagram shows the deployment of a model into a model registry:

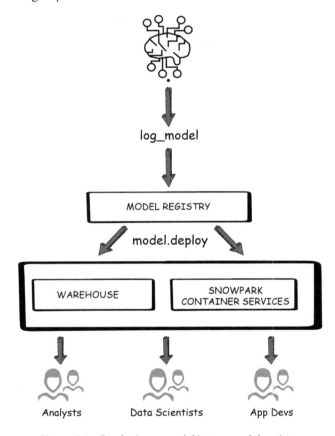

Figure 6.1 – Deploying a model into a model registry

The model registry is a Python API that manages models within the Snowflake environment, offering scalable, secure deployment and management capabilities for models within Snowflake. The Snowpark model registry is built upon a native Snowflake model entity, incorporating built-in versioning support for more streamlined management of models.

Preparing the model

To illustrate the model registration process, we'll efficiently craft a streamlined XGBoost model using minimal parameters, leveraging grid search on the *Bike Sharing* dataset. The BSD_TRAINING table prepared in the previous chapter is the foundational dataset for constructing our XGBoost model.

Here, we are making a feature list and finding label and output columns:

```
FEATURE_LIST = [ "HOLIDAY", "WORKINGDAY", "HUMIDITY", "TEMP", "ATEMP",
    "WINDSPEED", "SEASON", "WEATHER"]
LABEL_COLUMNS = ['COUNT']
OUTPUT_COLUMNS = ['PREDICTED_COUNT']
df = session.table("BSD_TRAINING")
df = df.drop("DATETIME","DATE")
df.show(2)
```

This will print out the following DataFrame:

"SEASON"	"HOLIDAY"	"WORKINGDAY"	"WEATHER"	"TEMP"	"ATEMP"	"HUMIDITY"	"WINDSPEED"	"CASUAL"	"REGISTERED"	"COUNT"	"HOUR"
1	0	0	1	9.84	14.395	81	12.7993954069447	3	13	16	0
1	0	0	1	9.02	13.635	80	12.7993954069447	8	32	40	1

Figure 6.2 – Model DataFrame

For the sake of simplicity, we will focus on optimizing two parameters within XGBoost:

```
from snowflake.ml.modeling.model_selection import GridSearchCV
from snowflake.ml.modeling.xgboost import XGBRegressor
param_grid = {
    "max_depth":[3, 4, 5, 6, 7, 8],
    "min_child_weight":[1, 2, 3, 4],
}
grid_search = GridSearchCV(
    estimator=XGBRegressor(),
    param_grid=param_grid,
    n_jobs = -1,
    scoring="neg_root_mean_squared_error",
    input_cols=FEATURE_LIST,
    label_cols=LABEL_COLUMNS,
    output_cols=OUTPUT_COLUMNS
)
```

This code employs Snowflake's ML module to perform a grid search for hyperparameter tuning on a gradient boosting regressor. It explores combinations of max_depth and min_child_weight within specified ranges, aiming to optimize the model based on the input and label columns provided.

The subsequent logical progression involves partitioning the dataset into training and testing sets:

```
train_df, test_df = df.random_split(weights=[0.7, 0.3], seed=0)
grid_search.fit(train_df)
```

This division is essential to facilitate model fitting, allowing us to train the model on the designated training dataset.

Extracting the optimum parameter

Having successfully trained our dataset using the XGBoost model, the next imperative is identifying optimal parameter values defined through grid search. Remarkably similar to the process in the scikit-learn package, Snowpark ML offers a comparable methodology. The ensuing code mirrors the steps in extracting these optimal parameters and subsequently visualizing them, demystifying the process for seamless comprehension:

```
import pandas as pd
import seaborn as sns
import matplotlib.pyplot as plt

gs_results = grid_search.to_sklearn().cv_results_
max_depth_val = []
min_child_weight_val = []
for param_dict in gs_results["params"]:
    max_depth_val.append(param_dict["max_depth"])
    min_child_weight_val.append(param_dict["min_child_weight"])
mape_val = gs_results["mean_test_score"]*-1
```

The preceding code uses pandas, seaborn, and matplotlib to analyze and visualize grid search results from a Snowpark ML model. It extracts the parameters, such as max_depth and min_child_weight, along with the corresponding **mean absolute percentage error** (**MAPE**) values for evaluation. The following code showcases the values:

```
gs_results_df = pd.DataFrame(data={
    "max_depth":max_depth_val,
    "min_child_weight":min_child_weight_val,
    "mape":mape_val})
sns.relplot(data=gs_results_df, x="min_child_weight",
    y="mape", hue="max_depth", kind="line")
```

The preceding code creates a pandas DataFrame named gs_results_df from listed max_depth, min_child_weight, and mape values. It then utilizes seaborn to generate a line plot, visualizing the relationship between learning rates, MAPE scores, and different numbers of estimators. Finally, the matplotlib plt.show() command displays the following plot:

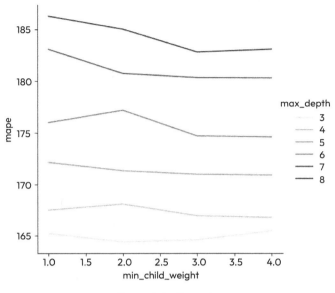

Figure 6.3 – Data plot

Upon careful observation of the previous plot, it becomes evident that a `max_depth` value of 8 paired with a `min_child_weight` learning rate of 2 yields the optimal results. It's noteworthy that, akin to `scikit-learn`, Snowpark ML offers streamlined methods for extracting these optimal parameters, simplifying the process for enhanced user convenience:

```
grid_search.to_sklearn().best_estimator_
```

The code transforms the Snowpark ML grid search results into a format compatible with `scikit-learn` and then retrieves the best estimator, representing the model with optimal hyperparameters:

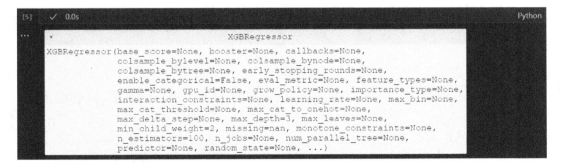

Figure 6.4 – Snowpark ML grid search result

In the next section, we will use the Snowpark model registry to log the model.

Logging the optimal model

With our optimal model in hand, the pivotal phase of the model registry unfolds. Much akin to the previously created model, we can extend this process to encompass multiple models, registering each through the model registry. In this case, we will be registering only our optimal model. We'll delve into a step-by-step exploration of how models can be registered and seamlessly deployed.

> **Note on the Model Registry and Feature Store**
>
> While we write this chapter, both the Model Registry and Feature Store are in private preview. Once they are open to all, the API methods might be slightly different from what we see in this book.

Next, we need to create a registry to log our model:

```
from snowflake.ml.registry import model_registry
registry = model_registry.ModelRegistry(session=session,
    database_name="SNOWPARK_DEFINITIVE_GUIDE",
    schema_name="MY_SCHEMA", create_if_not_exists=True)
```

The preceding code snippet imports the Snowflake ML model registry and initializes a `ModelRegistry` instance with session information, specified database, and schema names. If not existing, it creates a registry in the `SNOWPARK_DEFINITIVE_GUIDE` database and `MY_SCHEMA` schema.

The following code prepares the essential details to log a model in the model registry:

```
optimal_model = grid_search.to_sklearn().best_estimator_
optimal_max_depth = \
    grid_search.to_sklearn().best_estimator_.max_depth
optimal_min_child_weight = \
    grid_search.to_sklearn().best_estimator_.min_child_weight

optimal_mape = gs_results_df.loc[
    (gs_results_df['max_depth']==optimal_max_depth) &
    (gs_results_df['min_child_weight']== \
        optimal_min_child_weight), 'mape'].values[0]
```

It extracts the optimal model, as determined by the grid search, and retrieves specific hyperparameters such as `max_depth`, `min_child_weight`, and optimal parameter values.

Having completed all necessary steps for model registration, the preceding code seamlessly integrates the gathered information to register our optimal XGBoost model officially in the model registry:

```
model_name = "bike_model_xg_boost"
model_version = 1
X = train_df.select(FEATURE_LIST).limit(100)
```

```
registry.log_model( model_name=model_name,
                    model_version=model_version,
                    model=optimal_model,
                    sample_input_data=X,
                    options={"embed_local_ml_library": True, \
                             "relax": True})
registry.set_metric(model_name=model_name,
                    model_version=model_version,
                    metric_name="mean_abs_pct_err",
                    metric_value=optimal_mape)
```

The code assigns a name (bike_model_xg_boost) and version (1) to the model and logs it into the registry with associated details, including the sample input data and specific options. Additionally, it sets a custom metric, MAPE (mean_abs_pct_err), for the registered model with its corresponding value (optimal_mape). To verify successful registration, execute the following code:

```
registry.list_models().to_pandas()
```

This will confirm whether our XGBoost model and the gradient boost model (only the XGBoost model are steps shown here to avoid unnecessary repetition) are appropriately listed in the model registry:

NAME	OUTPUT_SPEC	RUNTIME_ENVIRONMENT_SPEC	TYPE	URI	VERSION
bike_model_xg_boost	None	None	xgboost sfc//SNOWPARK_DEFINITVE_GUIDE."My_SCHEMA".SNO...		1

Figure 6.5 – Model registered in the model registry

In the iterative journey of experimentation with diverse models and varied parameter configurations, we diligently register each model within the model registry through a structured methodology that ensures that each model, fine-tuned and optimized, is stored efficiently for future use. In the next section, we will deploy the model from the registry using Snowpark MLOps and predict its results.

Model deployment

In the preceding sections, we navigated the intricate landscape of deploying models through complex **user-defined functions** (UDFs) or stored procedures. However, the new Snowpark model registry simplifies the cumbersome process. It enhances the maintainability of models by providing a streamlined and standardized framework for handling predictive models in a production setting. This shift in methodology optimizes operational efficiency and aligns seamlessly with contemporary practices in the dynamic field of data science. A standard model deployment would follow this naming convention:

```
model_deployment_name = model_name + f"{model_version}" + "_UDF"
registry.deploy(model_name=model_name,
                model_version=model_version,
                deployment_name=model_deployment_name,
```

```
        target_method="predict",
        permanent=True,
        options={"relax_version": True})
```

The preceding snippet generates a unique name for deploying the model as a UDF. It then deploys the specified model version using the generated deployment name for the `predict` target method, ensuring permanence in the deployment. Additionally, it includes an option to relax version constraints during deployment. Just as we've showcased the catalog of registered models, an equivalent insight into deployed models can be obtained using the following line of code:

```
registry.list_deployments(model_name, model_version).to_pandas()
```

This functionality provides a comprehensive view of models transitioning from registration to deployment within the system:

Figure 6.6 – Bike model deployment

Now, let's leverage our deployed model to infer predictions for the test data and assess the accuracy of our predictions against actual outcomes:

```
model_ref = model_registry.ModelReference(
    registry=registry,
    model_name=model_name,
    model_version=model_version)
result_sdf = model_ref.predict(
    deployment_name=model_deployment_name,
    data=test_df)
result_sdf.show()
```

The code initiates a `ModelReference` object, linking to a specific model within the registry by referencing its name and version. Subsequently, it leverages this reference to predict the provided test data using the specified deployment, resulting in a Snowpark DataFrame (`result_sdf`). Finally, it displays the expected results through the `show()` method, as shown in the following screenshot:

Figure 6.7 – Model result DataFrame

Having observed a comprehensive cycle encompassing model development, registration, and deployment, it's noteworthy that this process is replicable for any model-building endeavor through the model registry. In the subsequent section, we will elucidate several beneficial methods inherent in the model registry, elevating its usability and augmenting the overall modeling experience. Now that we have deployed the model, we will look at other model registry methods.

Model registry methods

Beyond the functionality outlined for model deployment, the model registry extends its utility with several beneficial methods designed for effective model maintenance and housekeeping activities. In this section, we will explore a selection of these methods to enhance our understanding of their practical applications. We will start with model metrics.

Model metrics

Linking metrics to your model version is a pivotal feature within the model registry. This functionality serves as a fundamental aspect, providing a systematic means to gauge the performance of each model version distinctly. By associating metrics, users gain valuable insights into the efficacy of different iterations, facilitating informed decision-making based on the quantitative evaluation of model performance across various versions. It also helps in automating the pipeline, thereby retraining if model metrics drop below the threshold value. This deliberate metrics integration enriches the comprehensive model management capabilities and establishes a structured framework for ongoing model evaluation and refinement:

```
registry.set_metric(model_name=model_name,
                    model_version=model_version,
                    metric_name="mean_abs_pct_err",
                    metric_value=optimal_mape)
```

The preceding line sets a custom metric, `mean_abs_pct_err`, for a specific model version in the model registry, assigning the calculated MAPE value to quantify the model's performance. It enhances the model registry's ability to track and evaluate the effectiveness of different model versions:

```
registry.get_metric_value(model_name=model_name,
                          model_version=model_version,
                          metric_name="mean_abs_pct_err")
```

This will print the following output:

Figure 6.8 – MAPE value of mean_abs_pct_err

In addition to setting, we can retrieve the value of a specific custom metric, `mean_abs_pct_err`, associated with a particular model version from the model registry. It allows users to access and analyze quantitative performance metrics for practical model evaluation and comparison across different versions:

```
registry.get_metrics(model_name=model_name,
    model_version=model_version)
```

Much like retrieving a specific metric for a deployed model, an analogous approach allows us to access a comprehensive list of all associated metrics for a given deployed model. This facilitates a holistic understanding of the model's performance, providing a detailed overview of various metrics related to its evaluation and contributing to a thorough analysis of its effectiveness:

```
{'mean_abs_pct_err': 64.83345565795898}
```

Figure 6.9 – Metrics of mean_abs_pct_err

We can find the value of metrics from the model in the registry. In the next section, we will cover model tags and descriptions.

Model tags and descriptions

Setting a tag name and description for a deployed model is crucial for effective experiment tracking and documentation. Tags and descriptions provide context and insights into the model's purpose, configuration, and notable characteristics. This aids in maintaining a structured and informative record, enhancing reproducibility, and facilitating a more comprehensive analysis of experiment outcomes:

```
registry.set_tag(model_name=model_name,
                 model_version=model_version,
                 tag_name="usage",
                 tag_value="experiment")
registry.list_models().to_pandas()[["NAME", "TAGS"]]
```

The provided code first sets a tag named `stage` with the `experiment_1` value for a specific model version in the model registry. This tagging is a contextual marker for the model's purpose or usage. The subsequent line retrieves and displays, in a tabular format, the names of all models along with their associated tags, showcasing the tagged information for each model:

	NAME	TAGS
0	bike_model_gradient_boost	None
1	bike_model_xg_boost	{\n "usage": "experiment"\n}

Figure 6.10 – Model tags

Another noteworthy aspect is the flexibility to modify and remove tags as necessary, allowing for a dynamic adjustment of our experiment design. This capability empowers users to iteratively refine contextual information associated with a model, providing meaningful and evolving tags. The ability to alter and remove tags enhances experiment design flexibility. It ensures that the documentation and context surrounding models can adapt to changing insights and requirements throughout the experimentation lifecycle:

```
registry.remove_tag(model_name=model_name,
                    model_version=model_version,
                    tag_name="usage")
registry.list_models().to_pandas()[["NAME", "TAGS"]]
```

The provided code initiates the removal of a specific tag, named usage, from a particular model version within the model registry. Following this operation, the subsequent line retrieves and displays, in a tabular format, the names of all models along with their associated tags. This showcases the updated information after removing the specified tag, providing a comprehensive view of models and their altered tag configurations:

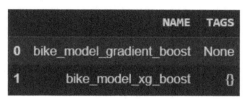

Figure 6.11 – Model tags removed

We can also provide descriptive information for a deployed model, offering valuable context and aiding future references. The ability to furnish a meaningful description enhances the comprehensibility of the model's purpose, configuration, or other pertinent details. The ensuing code block, which is self-explanatory and mirrors the process of setting tags, enables the assignment of a descriptive narrative to a deployed model, ensuring that vital information is encapsulated for reference in subsequent analyses or experiments:

```
registry.set_model_description(model_name=model_name,
    model_version=model_version,
    description="this is a test model")
print(registry.get_model_description(model_name=model_name,
    model_version=model_version))
```

The model description is set and can be retrieved to display on the screen:

```
WARNING:snowflake.snowpark:ModelRegistry.get_model_description() is in private preview since 0.2.0. Do not use it in production.
this is a test model
```

Figure 6.12 – Model description

Now that we have the model tags and description set, we will examine how to access the registry history.

Registry history

Accessing the registry history is an invaluable capability, offering a chronological account of model versions, associated metrics, tags, and descriptions. This historical perspective enhances transparency in model development and empowers data scientists to make informed decisions, track performance trends, and iterate on model improvements with precision. The ML registry, coupled with its history-tracking feature, thus emerges as a pivotal asset in the data science arsenal, fostering a structured and efficient approach to model development, deployment, and ongoing refinement:

```
registry.get_history().to_pandas()
```

The code retrieves and converts the entire history of the model registry into a pandas DataFrame, presenting a comprehensive tabular view of all recorded events and changes:

Figure 6.13 – Registry history

Narrowing down the search in the registry history is a common practice, and it can be achieved by specifying a model name and version. This targeted filtering allows for more focused exploration, aligning with typical preferences when navigating the model registry history:

```
registry.get_model_history(model_name=model_name,
    model_version=model_version).to_pandas()
```

This code fetches and converts the specific history of a particular model version, identified by its name and version, into a pandas DataFrame:

Figure 6.14 – Registry history filter

The resulting DataFrame offers a detailed chronological record of all events and changes associated with that specific model version within the registry. In the next section, we will learn about operations on the model registry.

Model registry operations

In the contemporary landscape of ML, the lifecycle of models is continually contracting, leading to shorter durations for deployed models. Concurrently, experiments with varying parameters generate many models, and their subsequent deployments are registered. This proliferation necessitates a thoughtful approach to model management, including periodic cleanup processes to maintain a streamlined and efficient model registry:

```
registry.delete_deployment(model_name=model_name,
    model_version=model_version,
    deployment_name=model_deployment_name)
    registry.list_deployments(model_name, model_version).to_pandas()
```

The preceding code deletes a specific deployment instance identified by the model's name, version, and deployment name from the model registry, ensuring efficient cleanup and management of deployed models:

MODEL_NAME	MODEL_VERSION	DEPLOYMENT_NAME	CREATION_TIME	TARGET_METHOD	TARGET_PLATFORM	SIGNATURE	OPTIONS	STAGE_PATH	ROLE

Figure 6.15 – Deleting a specific deployment

It serves as a method to remove obsolete or undesired deployments. We can also delete a whole model from the registry by using the following code:

```
registry.delete_model(model_name=model_name,
    model_version=model_version)
registry.list_models().to_pandas()
```

Similar to deleting a deployment, this code will delete a model from the model registry:

Figure 6.16 – Deleting a model

We can see that the entire model has been deleted from the registry. In the next section, we will look at the benefits of a model registry.

Benefits of the model registry in the model lifecycle

The Snowpark model registry streamlines the management of ML models throughout their lifecycle. Let's delve into how the model registry in Snowpark can assist in various stages of the ML model lifecycle:

1. **Model development**: During the development phase, data scientists can use Snowpark to build, train, and validate ML models directly within Snowflake. The model registry provides a centralized location to store and version control these models, making it easier to track changes, compare performance, and collaborate with team members.

2. **Model deployment**: Once a model is trained and validated, it needs to be deployed into production environments for inference. The model registry facilitates seamless deployment by providing a standardized interface to deploy models across different environments. This ensures consistency and reliability in model deployment processes.

3. **Model monitoring**: Monitoring the performance of deployed models is crucial for detecting drift and ensuring continued accuracy over time. The model registry can integrate with monitoring tools to track model performance metrics, such as accuracy, precision, recall, and F1-score, enabling proactive maintenance and optimization.

4. **Model governance**: Ensuring compliance with regulatory requirements and organizational policies is essential for responsible AI deployment. The model registry supports governance by providing capabilities for access control, audit logging, and versioning. This helps organizations maintain visibility and control over the entire model lifecycle.

5. **Model retraining and updating**: ML models need to be periodically retrained and updated to adapt to changing data distributions and business requirements. The model registry simplifies this process by enabling data scientists to seamlessly retrain models using updated data and algorithms while preserving the lineage and history of model versions.

6. **Model retirement**: As models become obsolete or are replaced by newer versions, they need to be retired gracefully. The model registry facilitates the retirement process by archiving outdated models, documenting reasons for retirement, and ensuring that relevant stakeholders are notified of changes.

The model registry offers an organized framework for model management and provides functionalities for efficient housekeeping, including setting and tracking metrics, tags, and descriptions. The registry's history-tracking capabilities have emerged as a valuable feature, allowing users to gain insights into the evolution of models over time. Tags and descriptions offer context and facilitate experiment tracking for accessing and filtering the registry history, enabling a comprehensive view of model-related activities. Overall, the model registry emerges as a powerful addition to Snowpark ML, centralizing model management, facilitating experimentation, and ensuring a streamlined and organized approach to model development and deployment.

Overall, the model registry in Snowpark plays a pivotal role in streamlining the ML model lifecycle, from development and deployment to monitoring, governance, retraining, and retirement. By providing a centralized platform for managing models, it helps organizations maximize the value of their ML investments while minimizing operational overhead and risks.

Managing Snowpark model data

In the previous section, we covered the deployment of ML models using the model registry. This section will look at managing Snowpark Models using feature stores. Snowpark ML Feature Store simplifies the feature engineering process and is integral to ML, significantly influencing model performance based on the quality of features employed. This chapter will help us learn about using feature stores and managing Snowpark models.

Snowpark Feature Store

The Snowpark Feature Store is an integrated solution for data scientists and ML engineers. It facilitates the creation, storage, management, and serving of ML features for model training and inference and is accessible through the Snowpark ML library. The feature store defines, manages, and retrieves features, supported by a managed infrastructure for feature metadata management and continuous feature processing. Its primary function is to make these features readily available for reuse in the ongoing development of future ML models. Feature stores play a pivotal role in operationalizing data input, tracking, and governance within the realm of feature engineering for ML:

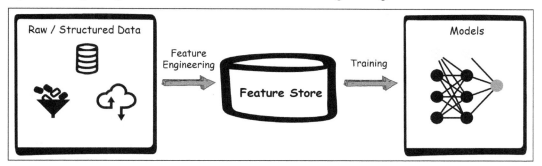

Figure 6.17 – Feature Store

By leveraging the Snowpark Feature Store, which is designed to simplify and enhance this process by offering increased efficiency for data scientists and ML practitioners. ML teams can uphold a singular and updated source of truth for model training, versioning, and inference features. We will use the *Bike Sharing* dataset and the ML model developed in the previous section to showcase how the Feature Store enhances the model development and deployment cycle.

Benefits of Feature Store

Utilizing feature stores provides several benefits for ML initiatives. Firstly, they enable feature reuse by saving developed features, allowing them to be quickly accessed and repurposed for new ML models, thereby saving time and effort. Secondly, they ensure feature consistency by providing a centralized registry for all ML features, maintaining consistent definitions and documentation across teams. Thirdly, feature stores help maintain peak model performance by centralizing feature pipelines, ensuring consistency between training and inference, and continuously monitoring data pipelines for any discrepancies.

Furthermore, feature stores enhance security and data governance by providing detailed information about each ML model's training data and deployment data, facilitating iteration and debugging. Integrating feature stores with cloud data warehouses enhances data security, ensuring the protection of both models and training data. Lastly, feature stores foster collaboration between teams by offering a centralized platform for the development, storage, modification, and sharing of ML features, promoting cross-team collaboration and idea-sharing for multiple business applications.

Feature stores versus data warehouses

Delving into the distinction between feature stores and data warehouses sheds light on their collaborative role in enhancing value within ML projects.

Similarities – shared traits and functions

Both feature stores and data warehouses exhibit parallels in their operational methodologies. They rely on **Extract, Transform, Load** (**ETL**) pipelines to facilitate data management and accessibility. Additionally, they serve as repositories endowed with metadata, fostering seamless data sharing and utilization across organizational teams.

End users – tailored utility

A notable deviation lies in their primary user base. Data warehouses traditionally cater to analysts entrenched in the generation of comprehensive business reports, delving into historical data for strategic insights. Conversely, feature stores cater specifically to data scientists immersed in the development of predictive ML models. While the latter may draw from data warehouses for supplementary insights, their core function revolves around leveraging feature stores for streamlined model development and inference.

Data types – structural variances

Structurally, data warehouses house domain-specific data within relational databases characterized by well-defined schemas. This structured format facilitates streamlined querying and retrieval of pertinent information, ideal for analytical endeavors. Conversely, feature stores house a distinct array of feature values crucial for ML model training. These values encompass both quantitative and categorical variables, enriching the model development process with granular insights.

ETL pipelines – divergent trajectories

The operational dynamics of ETL pipelines further accentuate the disparity between feature stores and data warehouses. ETL processes within data warehouses predominantly focus on data cleansing and transformation, ensuring data accuracy and coherence within the defined schema. In contrast, feature store pipelines embark on a more intricate journey, encompassing data extraction, transformation, and feature engineering. The transformations within feature stores often entail sophisticated computations and aggregations to distill intricate insights vital for model training and inference, underscoring their pivotal role in the ML lifecycle.

Now that we've grasped the essence of feature stores, comprehending their significance and differentiation from data warehouses, let's delve deeper into the various components comprising a feature store.

In the subsequent section, we'll embark on the creation of a rudimentary feature store tailored to the *Bike Sharing* dataset, focusing solely on weather-related features. The process entails the following:

1. Feature store creation
2. Feature entity creation

3. Selecting and transforming weather features

4. Creating a feature view

5. Generating datasets enriched with the feature view

6. Constructing an ML model empowered by the enriched dataset

7. Facilitating predictions based on the trained model

Let's discuss each of them in detail.

Creating a feature store

Initiating work with the Snowflake Feature Store involves establishing a new feature store or connecting to an existing one. This is accomplished by furnishing specific details to the `FeatureStore` constructor, including a Snowpark session, database name, feature store name, and default warehouse name. The `creation_mode` parameter is crucial in determining whether a new feature store should be created if it does not exist. To implement this functionality, we'll use the following code:

```
from snowflake.ml.feature_store import (
    FeatureStore, FeatureView, Entity, CreationMode)
fs = FeatureStore(
    session=session,
    database="SNOWPARK_DEFINITIVE_GUIDE",
    name="BIKE_SHARING_FEATURES",
    default_warehouse="COMPUTE_WH",
    creation_mode=CreationMode.CREATE_IF_NOT_EXIST,
)
```

This will open a session to the feature store and allow it to be accessed in the Snowpark session. The next step will be to set up a feature entity on this feature store.

Creating feature entities

Entities are fundamental elements linked with features and feature views, providing the cornerstone for feature lookups by defining join keys. Users can generate novel entities and formally register them within the feature store, thereby fostering connections and relationships between various features. This code creates an entity named WEATHER with an ID join key, registers it in the feature store (`fs`), and then displays a list of entities in the feature store:

```
entity = Entity(name="ENTITY_WEATHER", join_keys=["ID"])
fs.register_entity(entity)
fs.list_entities().show()
```

This generates the following output:

Figure 6.18 – Feature entity

The ENTITY_WEATHER entity has been created with the ID as the join key. The next step is to set up feature views.

Creating feature views

Within a feature store, feature views act as comprehensive pipelines, systematically transforming raw data into interconnected features at regular intervals. These feature views are materialized from designated source tables, ensuring incremental and efficient updates as fresh data is introduced. In our previous chapter, we explored a dataset that comprised various weather-related features. To preprocess this data effectively, we employed a Snowpark pipeline.

Through this pipeline, we performed transformations on the SEASON and WEATHER columns using one-hot encoding techniques. Additionally, we normalized the TEMP column to ensure consistency and facilitate model training. Given that we thoroughly discussed each step of this pipeline in our previous chapter, we'll be revisiting it briefly, focusing more on a high-level overview rather than delving into detailed explanations:

```
import snowflake.ml.modeling.preprocessing as snowml
from snowflake.ml.modeling.pipeline import Pipeline
from snowflake.snowpark.types import IntegerType

# CREATING ID COLUMN
from snowflake.snowpark.functions \
    import monotonically_increasing_id
df = df.withColumn("ID", monotonically_increasing_id())

df = df.drop("DATETIME","DATE")
CATEGORICAL_COLUMNS = ["SEASON","WEATHER"]
CATEGORICAL_COLUMNS_OHE = ["SEASON_OE","WEATHER_OE"]
MIN_MAX_COLUMNS = ["TEMP"]
import numpy as np
categories = {
    "SEASON": np.array([1,2,3,4]),
    "WEATHER": np.array([1,2,3,4]),
}
```

This code block utilizes Snowflake's ML capabilities for data preprocessing. It imports necessary modules such as preprocessing functions and the `Pipeline` class. The code creates a new `ID` column with unique identifiers for each row and drops unnecessary columns. It defines lists of categorical columns and their transformed versions after one-hot encoding, along with columns to be normalized. Additionally, it specifies categories for each categorical column, likely for encoding purposes, facilitating effective ML model processing:

```
preprocessing_pipeline = Pipeline(
    steps=[
        (
            "OE",
            snowml.OrdinalEncoder(
                input_cols=CATEGORICAL_COLUMNS,
                output_cols=CATEGORICAL_COLUMNS_OHE,
                categories=categories
            )
        ),
        (
            "MMS",
            snowml.MinMaxScaler(
                clip=True,
                input_cols=MIN_MAX_COLUMNS,
                output_cols=MIN_MAX_COLUMNS,
            )
        )
    ]
)
transformed_df = preprocessing_pipeline.fit(df).transform(df)
transformed_df.show()
```

In the first step, an ordinal encoder (OE) is applied to transform categorical columns specified in the CATEGORICAL_COLUMNS list into their one-hot encoded versions, as defined by the CATEGORICAL_COLUMNS_OHE list. The categories parameter specifies the categories for each categorical column, likely used for encoding purposes.

In the second step, a min-max scaler (MMS) is used to normalize columns specified in the MIN_MAX_COLUMNS list. This scaler ensures that values in these columns are scaled to a specific range, typically between 0 and 1, while preserving their relative proportions.

The preprocessing pipeline is then applied to the df DataFrame using the fit-transform paradigm, where the pipeline is first fit to the data to learn parameters (for example, category mappings for ordinal encoding), and then applied to transform the DataFrame. The transformed DataFrame is then displayed using the show() method. Overall, this code prepares the DataFrame for further analysis or model training by preprocessing its columns using the specified pipeline. The resultant DataFrame is as follows:

"TEMP"	"SEASON_OE"	"WEATHER_OE"	"SEASON"	"HOLIDAY"	"WORKINGDAY"	"WEATHER"	"ATEMP"	"HUMIDITY"	"WINDSPEED"
0.22448979591836735	0.0	0.0	1	0	0	1	14.395	81	0.0
0.20408163265306123	0.0	0.0	1	0	0	1	13.635	80	0.0
0.20408163265306123	0.0	0.0	1	0	0	1	13.635	80	0.0
0.22448979591836735	0.0	0.0	1	0	0	1	14.395	75	0.0
0.22448979591836735	0.0	0.0	1	0	0	1	14.395	75	0.0
0.22448979591836735	0.0	1.0	1	0	0	2	12.88	75	6.0032
0.20408163265306123	0.0	0.0	1	0	0	1	13.635	80	0.0
0.18367346938775508	0.0	0.0	1	0	0	1	12.88	86	0.0
0.22448979591836735	0.0	0.0	1	0	0	1	14.395	75	0.0
0.3061224489795918	0.0	0.0	1	0	0	1	17.425	76	0.0

Figure 6.19 – Transformed DataFrame

Throughout the model-building process, various models are constructed using subsets of features, such as weather features and time-related features. Additionally, models are developed using combined data to ascertain superior performance. To expedite the model-building process and reduce data engineering overheads, we'll organize weather-related features into a dedicated feature view. Subsequently, we'll leverage this feature view to generate datasets and construct an XGBoost model in the ensuing section:

```
feature_df = transformed_df.select(["SEASON_OE",
    "WEATHER_OE", "TEMP", "ATEMP", "HUMIDITY",
    "WINDSPEED", "ID"])

fv = FeatureView(
    name="WEATHER_FEATURES",
    entities=[entity],
    feature_df=feature_df,
    desc="weather features"
)

fv = fs.register_feature_view(
    feature_view=fv,
    version="V1",
    block=True
)
fs.read_feature_view(fv).show()
```

The code selects specific columns from the DataFrame to create a feature DataFrame (feature_df). Then, it constructs a feature view named WEATHER_FEATURES associated with the previously defined entity and registers it in the feature store (fs) with version V1. The resulting DataFrame is as follows:

```
|"SEASON_OE" |"WEATHER_OE" |"TEMP"                |"ATEMP" |"HUMIDITY" |"WINDSPEED" |"ID" |
---------------------------------------------------------------------------------------------
|0.0         |0.0          |0.22448979591836735   |14.395  |81         |0.0         |0    |
|0.0         |0.0          |0.20408163265306123   |13.635  |80         |0.0         |1    |
|0.0         |0.0          |0.20408163265306123   |13.635  |80         |0.0         |2    |
|0.0         |0.0          |0.22448979591836735   |14.395  |75         |0.0         |3    |
|0.0         |0.0          |0.22448979591836735   |14.395  |75         |0.0         |4    |
|0.0         |1.0          |0.22448979591836735   |12.88   |75         |6.0032      |5    |
|0.0         |0.0          |0.20408163265306123   |13.635  |80         |0.0         |6    |
|0.0         |0.0          |0.18367346938775508   |12.88   |86         |0.0         |7    |
|0.0         |0.0          |0.22448979591836735   |14.395  |75         |0.0         |8    |
|0.0         |0.0          |0.3061224489795918    |17.425  |76         |0.0         |9    |
---------------------------------------------------------------------------------------------
```

Figure 6.20 – Feature DataFrame

Once developed, features can be systematically stored in the feature store, fostering their availability for reuse or seamless sharing among various ML models and teams. This functionality significantly accelerates the creation of new ML models, eliminating the redundancy of building each feature from scratch. In the same way, we can create another feature view as rental features by combining similar features.

Preparing the dataset

Once our feature pipelines are meticulously configured and ready, we can initiate their deployment to generate training data. Subsequently, these feature pipelines become instrumental in facilitating model prediction, marking the seamless transition from feature engineering to the practical application of ML models:

```
#GENERATING TRAINING DATA
spine_df = session.table("BSD_TRAINING")
spine_df = spine_df.withColumn("ID",
    monotonically_increasing_id())
spine_df = spine_df.select("ID", "COUNT")
spine_df.show()


train_data = fs.generate_dataset(
    spine_df=spine_df,
    features=[
        fv.slice([
            "HUMIDITY","SEASON_OE","TEMP",
            "WEATHER_OE","WINDSPEED"
        ])
    ],
    materialized_table=None,
    spine_timestamp_col=None,
    spine_label_cols=["COUNT"],
    save_mode="merge",
```

```
    exclude_columns=['ID']
)


train_data.df.show()
```

Creating training data becomes straightforward as materialized feature views inherently encompass crucial metadata such as join keys and timestamps for **point-in-time** (**PIT**) lookup:

```
|"COUNT"  |"HUMIDITY"  |"SEASON_OE"  |"TEMP"               |"WEATHER_OE"  |"WINDSPEED"  |
----------------------------------------------------------------------------------------
|16       |81          |0.0          |0.22448979591836735  |0.0           |0.0          |
|40       |80          |0.0          |0.20408163265306123  |0.0           |0.0          |
|32       |80          |0.0          |0.20408163265306123  |0.0           |0.0          |
|13       |75          |0.0          |0.22448979591836735  |0.0           |0.0          |
|1        |75          |0.0          |0.22448979591836735  |0.0           |0.0          |
|1        |75          |0.0          |0.22448979591836735  |1.0           |6.0032       |
|2        |80          |0.0          |0.20408163265306123  |0.0           |0.0          |
...
|8        |75          |0.0          |0.22448979591836735  |0.0           |0.0          |
|14       |76          |0.0          |0.3061224489795918   |0.0           |0.0          |
```

Figure 6.21 – Training data

The process primarily involves supplying spine data—termed so because it serves as the foundational structure enriched by feature joins. In our case, spine data encompasses the feature to be predicted—COUNT—along with the join key column ID. Moreover, the flexibility to generate datasets with subsets of features within the feature view is available through slicing. Now that we have the training data ready, we will use it to train the model and predict the data output using the feature store.

The preparation of all data—both for training and operational use—requires meticulous handling through feature pipelines. These pipelines, resembling traditional data pipelines, aggregate, validate, and transform data output in a format suitable for input into the ML model. Properly orchestrated feature pipelines ensure that data is refined before being fed into the model, maintaining the integrity and relevance of features derived from the training process.

Model training

We covered the model-building process extensively in the previous chapter, so in this section, we will focus on building it using the training dataset generated from feature views from the feature store. We are using a similar method outlined in the previous chapter in training a gradient boost model but just using feature views:

```
from snowflake.ml.modeling.model_selection import GridSearchCV
from snowflake.ml.modeling.ensemble \
    import GradientBoostingRegressor
```

```
FEATURE_LIST = ["TEMP", "WINDSPEED", "SEASON_OE", "WEATHER_OE"]
LABEL_COLUMNS = ['COUNT']
OUTPUT_COLUMNS = ['PREDICTED_COUNT']

param_grid = {
    "n_estimators":[100, 200, 300, 400, 500],
    "learning_rate":[0.1, 0.2, 0.3, 0.4, 0.5],
}

grid_search = GridSearchCV(
    estimator=GradientBoostingRegressor(),
    param_grid=param_grid,
    n_jobs = -1,
    scoring="neg_root_mean_squared_error",
    input_cols=FEATURE_LIST,
    label_cols=LABEL_COLUMNS,
    output_cols=OUTPUT_COLUMNS
)
train_df = train_data.df.drop(["ID"])
grid_search.fit(train_df)
```

The elegance of Snowpark surfaces in this simplicity, as no significant modifications are needed to train a model using feature views seamlessly. We will create testing data to test the model for accuracy using the following code:

```
test_df = spine_df.limit(3).select("ID")
enriched_df = fs.retrieve_feature_values(
    test_df, train_data.load_features())
enriched_df = enriched_df.drop('ID')
enriched_df.show()
```

This creates a test DataFrame (`test_df`) by selecting the ID column from the first three rows of `spine_df`. Then, it retrieves and displays feature values for the test data frame using the feature store and training data generated from feature views:

"HUMIDITY"	"SEASON_OE"	"TEMP"	"WEATHER_OE"	"WINDSPEED"
81	0.0	0.22448979591836735	0.0	0.0
80	0.0	0.20408163265306123	0.0	0.0
80	0.0	0.20408163265306123	0.0	0.0

Figure 6.22 – Testing data

Now that the testing data is ready, we can predict the model using this data to get the results.

Model prediction

In this section, we will use the testing data generated from the feature store to make a prediction:

```
pred = grid_search.predict(enriched_df.to_pandas())
pred.head()
```

The prediction displays the results with the predicted count value, showing the number of customers using shared bikes at the given hour:

..	HUMIDITY	SEASON_OE	TEMP	WEATHER_OE	WINDSPEED	PREDICTED_COUNT
0	81	0.0	0.224490	0.0	0.0	70.897276
1	80	0.0	0.204082	0.0	0.0	62.313673
2	80	0.0	0.204082	0.0	0.0	62.313673

Figure 6.23 – Model prediction

This shows how easy and improved it is to use a feature store to build a Snowpark ML model. In the next section, we will highlight some benefits of using feature stores.

When to utilize versus when to avoid feature stores

Feature stores are particularly advantageous in ML processes when there's a need for efficient feature management and reuse across multiple models or teams. They shine in the following scenarios:

- **Feature reuse**: Features need to be reused or shared between different ML models or teams, reducing redundant efforts in feature engineering

- **Consistency and governance**: Ensuring consistent definitions, documentation, and governance of features across diverse ML projects or teams is critical

- **Model performance**: Maintaining peak model performance by ensuring consistency between feature definitions in training and inference pipelines, thus avoiding performance degradation due to discrepancies

- **Data collaboration**: Fostering collaboration between different teams or stakeholders involved in ML projects by offering a centralized platform for feature development, storage, modification, and sharing

- **Scalability**: Handling large volumes of features and data efficiently, especially in environments where data is continuously evolving or being updated

However, feature stores may not be necessary in the following scenarios:

- **Simple models**: For simple models with few features and minimal complexity, the overhead of setting up and maintaining a feature store may outweigh its benefits

- **Static data**: In cases where the data is relatively static and doesn't require frequent updates or feature engineering, the need for a feature store may be limited

- **Limited collaboration**: When ML projects involve a small, tightly-knit team working on isolated tasks without the need for extensive collaboration or feature sharing, the use of a feature store may be unnecessary

- **Resource constraints**: Organizations with limited resources or infrastructure may find it challenging to implement and maintain a feature store effectively

In summary, while feature stores offer numerous benefits for efficient feature management in ML projects, their adoption should be carefully considered based on the specific needs and constraints of each project or organization.

Summary

This chapter discussed the model registry and the importance of meaningful tags and descriptions, offering context and facilitating experiment tracking. We also highlighted different methods of operating with the model registry. We navigated through the capabilities of the Snowflake Feature Store within the Snowpark ML ecosystem and how to utilize it for managing Snowflake models.

In the next chapter, we will learn about developing native applications using the Snowpark framework.

Part 3: Snowpark Applications

This part focuses on developing and deploying apps and LLMs on Snowflake using the **Native App Framework** and **Container Services**.

This part includes the following chapters:

- *Chapter 7, Developing a Native Application with Snowpark*
- *Chapter 8, Introduction to Snowpark Container Services*

7

Developing a Native Application with Snowpark

In today's data-driven world, the demand for actionable insights is at an all-time high. However, traditional data analysis workflows often need help with slow processing times, siloed data, and complex development procedures. This is where Snowpark, with its unique features for native applications, comes to the rescue. By offering a powerful and innovative solution, Snowpark overcomes these challenges and opens up new revenue streams for organizations through distribution in the Snowflake Marketplace.

In this chapter, we're going to cover the following main topics:

- Introduction to the Native App Framework
- Developing the native application
- Publishing the native application
- Managing the native application

Technical requirements

To set up the environment, please refer to the technical requirements in the previous chapter. The supporting materials are available at `https://github.com/PacktPublishing/The-Ultimate-Guide-To-Snowpark`.

Introduction to the Native Apps Framework

Data analysis has evolved beyond static reports. In today's business landscape, real-time insights, interactive visuals, and seamless integration with operational tools are in high demand. While traditional methods utilizing SQL or pre-built dashboards have their place, they need more agility and flexibility for modern data exploration. This is where native applications step in, offering exciting potential for immediate and actionable insights and sparking new possibilities for your data analysis.

Snowflake native applications offer a unique advantage by allowing you to develop applications directly within the Data Cloud. These applications reside alongside your data, eliminating the need for data movement or external connectors. This removes the data transfer bottleneck between systems and significantly reduces network latency. Snowflake's architecture, optimized for parallel processing and distributed storage, enables faster analysis of complex queries and large datasets than traditional models. Moreover, the native applications benefit from Snowflake's secure foundation, which offers a multi-layered security architecture with features such as encryption, access controls, and audit trails without separate infrastructure.

Snowflake's native application Landscape

The Snowflake Native App Framework is a powerful tool for building data applications that utilize Snowflake's core functionality. With this framework, you can extend the capabilities of Snowflake features by sharing data and relevant business logic with other Snowflake accounts. This can include using Streamlit apps, stored procedures, and functions written in the Snowpark API, JavaScript, and SQL. By leveraging the power of Snowflake's capabilities, you can create sophisticated applications that are tailored to your unique needs.

In addition to expanding Snowflake's features, the Snowflake Native App Framework also allows you to share your applications with others. You can distribute your apps through free or paid listings in the Snowflake Marketplace, or privately distribute them to specific consumers. Furthermore, this framework allows you to create rich visualizations in your application using Streamlit. With the Snowflake Native App Framework, you have all the tools you need to create powerful, scalable, and flexible data applications that meet your organization's needs.

In summary, the Snowflake Native App Framework provides a comprehensive solution for building sophisticated data applications. Whether you need to expand Snowflake's capabilities, share your applications with others, or create rich visualizations, this framework has you covered. With its powerful features and scalability, the Snowflake Native App Framework is an essential tool for any organization looking to build data applications that leverage Snowflake's full potential.

Snowflake native applications consist of business logic that defines the core functionality of your application, encompassing tasks such as data transformations, calculations, and model predictions. The data model can be developed using Snowpark, where native applications leverage Snowflake's powerful engine to interact with your data efficiently. Snowpark enables you to manipulate tables, query datasets, and extract meaningful insights. The user interface can be developed using Streamlit, and users can interact with the application natively within Snowflake.

Snowpark bridges the gap between traditional data analysis and modern application development by offering several key features that make it the ideal engine for building native applications. The applications seamlessly scale with your data volume, eliminating the need for manual infrastructure provisioning.

Before diving deep into developing a Snowpark application, let us understand the benefits that Native Apps provide.

Native App Framework components

The **Snowflake Native App Framework** components consist of the **provider**, the developer of the apps and the distributor of them for consumption, and the **consumer**, who accesses the data and application that the provider distributes. The high-level components of the Native App Framework are shown in the following diagram:

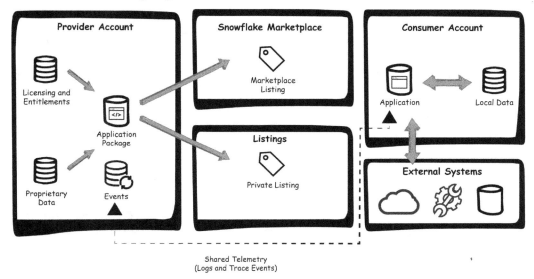

Figure 7.1 – Native App Framework components

The provider account is where the application is developed and packaged for distribution. The package includes the application code, Snowpark models, and associated data. It also consists of the various licenses and entitlements required for the application packages. The application is then packed and can be made available for distribution.

The provider can distribute the packages as a public marketplace listing that other organizations can consume or through a private listing, which can also be distributed as internal apps within the same organization. The consumer account deploys the application and can consume it with local data and access external systems. The app for this can be built using Streamlit in Python.

Streamlit in Snowflake

Streamlit is an open source Python library that makes creating and sharing custom web apps for machine learning and data science easy. It allows you to build and deploy powerful data applications quickly. Streamlit in Snowflake helps you move beyond pure data and enter the territory of application logic, making the application come alive and transforming insights into actions and reactions. Using Streamlit in Snowflake, you can build applications that process and use data in Snowflake without moving data or application code to an external system. The Native App can embed interactive data visualizations and user interfaces directly into the app, enabling dynamic charts and graphs that update as users interact with the data, and interactive data exploration tools for users to drill down and uncover more profound insights.

Let's discuss the various benefits provided by Native Apps.

Benefits of Native Apps

Native applications built within Snowflake offer unparalleled advantages for data analysis and decision-making. Native Apps provide multiple benefits; by offering streamlined development with Snowpark, they eliminate the steep learning curve typically associated with a development framework. Developers can seamlessly transition to data manipulation within Snowflake's robust cloud environment, allowing rapid innovation. This empowers developers to focus on building impactful applications and extracting valuable insights without syntax roadblocks.

Snowflake's integration with Streamlit enables developers to unleash the power of their existing Python expertise directly within Snowflake. Developers can build sophisticated applications for data exploration, machine learning, and interactive dashboards. This eliminates context switching and maximizes productivity within Snowflake's secure and scalable environment. Snowflake's *data residency* principle ensures that valuable information remains within the secure perimeter, significantly reducing the risk of data breaches.

Native applications seamlessly interface with existing Snowflake features and tools, such as Tasks and Streams, creating an integrated analytics environment. Leveraging Snowflake Tasks for automated execution and scheduling, along with Snowflake Streams or Snowpipe for real-time data ingestion and delivery, ensures continuous updates and enhances performance. Snowflake **user-defined functions** (**UDFs**) and stored procedures also improve flexibility and minimize data movement. We will explore Streamlit in Snowflake in the next section.

As we delve deeper into this chapter, we will utilize the familiar Bike Sharing dataset and its associated Snowpark model we saw in the previous chapters and develop a Streamlit application using Snowsight.

Developing the native application

The initial step involves developing a Streamlit application to illustrate the development of the Snowpark model. The application provides a user-friendly interface that streamlines the exploration and understanding of Snowpark's underlying data and modeling techniques. Within Snowflake, the Streamlit application can be developed through a dedicated Python editor within Snowsight. This editor empowers users to write, edit, and execute code for their Streamlit applications effortlessly. Notably, it offers invaluable features such as auto-completion and comprehensive documentation, facilitating a smooth and intuitive development experience while leveraging the full potential of Streamlit and Snowpark functionalities. We will be using this editor to develop the Streamlit application.

Let's delve into the detailed steps of creating a Streamlit application within the Snowflake environment:

> **Note**
>
> The **CREATE STREAMLIT** privilege is required for the user at the schema level to develop the application and this application will utilize warehouse compute to run it, so choose the warehouse size appropriately. X-Small will be sufficient for the Streamlit and Native Apps developed in this chapter.

1. Upon successful login, select **Projects** from the left navigation bar within Snowsight and select **Streamlit** to initiate the Streamlit application:

Figure 7.2 – Streamlit Snowsight

2. Within the Streamlit projects section, click the **+ Streamlit App** button to create a new Streamlit application:

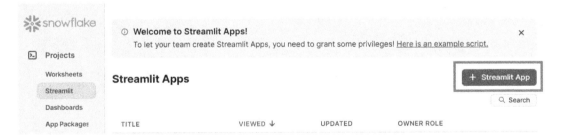

Figure 7.3 – Streamlit app creation

3. Once you click the button, the **Create Streamlit App** window will appear. Provide a name for your application, as shown here:

Figure 7.4 – Configuring the application details

4. Select the appropriate warehouse from the drop-down menu where you wish to execute your application's queries. Specify the database and schema where the application will be hosted from the respective drop-down menus:

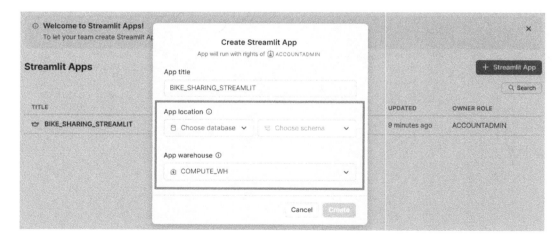

Figure 7.5 – Defining the execution environment

5. Finalize the creation process by clicking the **Create** button, which initiates the setup of your Streamlit application. Upon successful creation, Streamlit will be launched in your Snowflake editor, presenting an example of the Streamlit application in the **Viewer** mode. This mode offers a preview of how the application will appear to end users, facilitating visualization and assessment of its user interface:

Figure 7.6 – The Streamlit Snowflake editor

In the next section, we will discover and familiarize ourselves with the Streamlit editor.

The Streamlit editor

The Streamlit editor in Snowflake helps develop the Streamlit application within Snowflake. The editor is built into the Snowsight interface, making it easier to create without needing an external environment. The Streamlit in Snowflake interface is structured into three primary panes.

The object browser provides visibility into the databases, schemas, and views accessible under your permissions, aiding in data exploration and access. The Streamlit editor houses a dedicated Python editor tailored to craft your Streamlit application code, facilitating seamless development and customization. The Streamlit preview displays the running Streamlit application in real time, allowing you to observe its behavior and appearance as you make modifications.

By default, only the Streamlit editor and preview panes are visible. However, you can adjust the interface layout according to your preferences. Use the **Show/Hide** buttons in the lower-left corner of the Streamlit in Snowflake editor to toggle the visibility of different panes according to your workflow requirements.

Now that we are familiar with the interface, let's create our first Streamlit application in the next section.

Running the Streamlit application

In this section, we will build an application using the Streamlit code from the chapter_7 folder of the GitHub repository (https://github.com/PacktPublishing/The-Ultimate-Guide-To-Snowpark/tree/main/chapter_7). The SnowparkML code has been extensively covered in previous chapters and is available in the repository. Copy and paste the code in the streamlit_bike_share_analysis.py file from the repository into the Streamlit editor. The code requires the BSD_TRAIN table created from the *Bike Sharing* dataset that we have used before. The chapter_7_data_load.ipynb file can be run to create the required table.

Now, execute the Streamlit application by selecting the **Run** option to refresh, and the result will be updated within the Streamlit preview pane:

Figure 7.7 – The Streamlit App preview

The code requires several packages to be installed in your Streamlit environment. Click on the **Packages** button in the code editor and select the following packages:

- Matplotlib
- Seaborn
- SciPy
- Snowflake-ml-python

You have successfully run your first Streamlit app inside Snowflake. You can select the app to execute it within the Snowsight context:

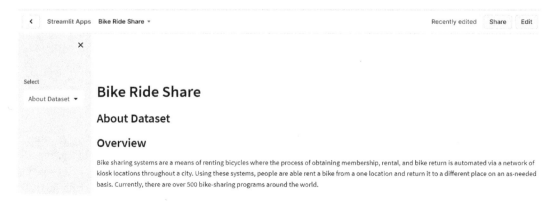

Figure 7.8 – The Streamlit app

> **Note**
>
> Snowsight offers a convenient platform for crafting single-page Streamlit applications. However, deploying multi-page Streamlit apps requires a slightly different approach, necessitating the utilization of SQL commands. Once the deployment process is complete, you can seamlessly navigate and interact with the multi-page application directly within Snowsight's interface.

In the following section, we will develop and push the Streamlit application using the Native App Framework.

Developing with the Native App Framework

Now that our Streamlit application is operational, let's explore the process of transforming it into a Native App within the Snowflake ecosystem. In this section, we'll give a detailed walkthrough on leveraging the Native App Framework to develop applications to share data and associated business logic with other Snowflake accounts. We will primarily demonstrate this using the Snowsight web interface, which can also be created using a client such as VS Code.

The following sequential steps will convert your existing Streamlit app into a native application in Snowflake:

1. Creating the application files
2. Creating an application package
3. Uploading application files to the named stage
4. Installing the application
5. Versioning the application
6. Testing the application

We will start with the first step, creating the application files.

Creating the application files

Creating the application files involves creating the setup script and the manifest file, which are vital components that the Native App Framework requires.

Creating the setup script

The setup script is an SQL file that executes automatically when consumers install the application in their accounts. It facilitates seamless configuration and setup processes and consists of the initialization instructions required to configure the application. To initiate the creation of the setup script, follow these steps:

1. We'll start by creating a directory structure on your local filesystem. Create a folder named `native_apps` to serve as the root directory for your application's external files.

2. Within this folder, create a subfolder called `scripts` to store the script files. This folder will contain all the script files necessary for the application.

3. Inside the `scripts` folder, create a new file called `setup.sql` and add the following statement:

```sql
-- Setup script for the Bike Share Streamlit application.
CREATE APPLICATION ROLE app_public;
CREATE OR ALTER VERSIONED SCHEMA code_schema;
GRANT USAGE ON SCHEMA code_schema
  TO APPLICATION ROLE app_public;
CREATE STREAMLIT code_schema.bike_share_streamlit
  FROM '/streamlit'
  MAIN_FILE = '/streamlit_bike_share_analysis.py';
GRANT USAGE ON STREAMLIT code_schema.bike_share_streamlit
  TO APPLICATION ROLE app_public;
```

We will be seeing the detailed explanation of the previous code as we move through this section.

Creating the Streamlit app file

We will incorporate the application logic and the Streamlit app, leveraging its capabilities for data science and machine learning tasks. Utilizing the Streamlit app enhances user interaction and facilitates data visualization. We will start the process by creating the Streamlit app:

1. Create a subfolder named `streamlit` within the `native_apps` directory.

2. Within the folder, create a file called `streamlit_bike_share_analysis.py`. The code for this is available within the repository inside the `chapter_7` folder.

3. Copy and paste the following code into this file and save it:

```python
# Set page config
st.set_page_config(layout="wide")
# Get current session
session = get_active_session()
@st.cache_data()
def load_data():
    # Load Bike Share data
    snow_df = session.table(
        "SNOWPARK_DEFINITIVE_GUIDE.MY_SCHEMA.BSD_TRAIN")
```

For the sake of brevity, only a snippet of the Streamlit app file is shown here; refer to the `streamlit_bike_share_analysis.py` file under the `chapter_7` folder in the book GitHub repository for the complete code.

Creating the README file

The README file serves as a descriptive guide outlining your application's functionality. It is readily accessible when viewing your application within Snowsight. To generate the README file for your application, follow these steps:

1. Navigate to the `native_apps` directory and create a file called `readme.md`.

2. Populate the `readme.md` file with the following content detailing the purpose and features of your application:

    ```
    # Bike Sharing Analysis - Through Native Apps.
    This native app presents the analysis we performed on the Kaggle
    dataset - Bike sharing demand as an interactive dashboard
    ```

Now we can create the manifest file in the next section.

Creating the manifest file

The manifest file, which uses YAML syntax, contains essential configuration details about the application. It is a foundational document outlining crucial specifications for proper functionality, and every application developed within the Native App Framework necessitates a corresponding manifest file.

To generate the manifest file, proceed as follows:

1. Within the `native_apps` folder, create a new file named `manifest.yml`.

2. Populate the `manifest.yml` file with the following content:

    ```
    manifest_version: 1
    artifacts:
      setup_script: scripts/setup.sql
      readme: readme.md
    ```

The `setup_script` property specifies the location of the setup script relative to the manifest file's location. Ensure that the path and filename specified correspond precisely to the setup script created earlier.

> **Note**
>
> The manifest file is crucial for your project. It must be named `manifest.yml` and placed at the root level. All file paths, including the setup scripts, are relative to this file's location. Fundamental properties such as `manifest_version`, `artifacts`, and `setup_script` are mandatory to ensure proper functioning. You can optionally include the `readme` property to provide additional documentation.

In the next section, we will create an application package for the native apps.

Creating an application package

In this section, you will create an application package, a container for the application's resources. The application package extends Snowflake's database to encompass additional application-related information, serving as a consolidated container housing shared data content and applications. To create an application package, your role needs to have the **Create Application Package** privilege. Execute the following command to grant the privilege:

```
GRANT CREATE APPLICATION PACKAGE ON ACCOUNT
    TO ROLE accountadmin;
```

Once executed, you should see the following output:

Figure 7.9 – Role creation

Now, run the following command to create the SNOWFLAKE_NATIVE_APPS_PACKAGE application package:

```
CREATE APPLICATION PACKAGE snowflake_native_apps_package;
```

You will receive the following output that confirms the package has been created:

Figure 7.10 – Package creation

The application package will be created, and the current context will be switched to SNOWFLAKE_NATIVE_APPS_PACKAGE. We can confirm the list of packages by running the following command:

```
SHOW APPLICATION PACKAGES;
```

This will display the following output:

Figure 7.11 – Showing the packages

Now that the application packages are created, the next step is to make the named stage for uploading the application files. Switch to the context of the application package by running the following command:

```
USE APPLICATION PACKAGE snowflake_native_apps_package;
```

This will display the following output:

Figure 7.12 – Using snowflake_native_apps_package

We will need a schema to host the stage. Create a schema by running the following command:

```
CREATE SCHEMA my_schema;
```

Once the schema is created, you'll see the following message:

Figure 7.13 – Creating a new schema

The next step is to create the stage in this schema. Create a stage by running the following code:

```
CREATE OR REPLACE STAGE snowflake_native_apps_package.my_schema.my_
stage FILE_FORMAT = (TYPE = 'csv' FIELD_DELIMITER = '|' SKIP_HEADER =
1);
```

With this step, you've established a named stage within the application package, providing a designated space to upload the files essential for building your application.

> **Note on setup.sql**
>
> The following set of SQL commands are part of `setup.sql` and are packaged inside the native apps folder in `chapter_7`. This file will be uploaded to the stage area and will be utilized when we create a Native Apps application, so the code need not be executed in the Snowflake worksheet.

Let us see what the `set.sql` file encompasses to create a Native Apps application. We will also be creating a role that is required for the application. To create an application role, run the following command:

```
CREATE APPLICATION ROLE app_public;
CREATE OR ALTER VERSIONED SCHEMA code_schema;
GRANT USAGE ON SCHEMA code_schema TO APPLICATION ROLE app_public;
```

This code initializes an application role named `app_public`. It then creates or updates a versioned schema named `code_schema`. Finally, it grants permission for the `app_public` application role to use the `code_schema` schema. This setup is commonly used for managing access control and schema organization in database applications.

Run the following code to incorporate the Streamlit app:

```
CREATE STREAMLIT code_schema.bike_share_streamlit
  FROM '/streamlit'
  MAIN_FILE = '/streamlit_bike_share_analysis.py';

GRANT USAGE ON STREAMLIT code_schema.bike_share_streamlit
  TO APPLICATION ROLE app_public;
```

This code creates a Streamlit app named `bike_share_streamlit` within the `code_schema` schema, pulling its content from the `/streamlit` directory and its main file from `/streamlit_bike_share_analysis.py`. It then grants usage permission for this Streamlit app to the `app_public` application role, enabling access for authorized users within the application's scope. This facilitates integrating a Streamlit-based bike share analysis tool into the database setup, with controlled access managed by the application role.

Uploading application files to the named stage

The application files need to be uploaded to the named stage. They can be uploaded via Snowsight or through Snowpark Python code, as described in *Chapter 2*, Establishing a Foundation with Snowpark. For readers using the SnowSQL client to upload the files to the stage, perform the following steps:

1. Navigate to the `native_apps` folder on your local machine.

2. Run the following SnowSQL commands to upload all folders and files:

```
PUT file:///<path_to_your_root_folder>/native_apps/manifest.yml
@snowflake_native_apps_package.my_schema.my_stage overwrite=true
auto_compress=false;
PUT file:///<path_to_your_root_folder>/native_apps/scripts/
setup.sql @snowflake_native_apps_package.my_schema.my_stage/
scripts overwrite=true auto_compress=false;
PUT file:///<path_to_your_root_folder>/native_apps/streamlit/
streamlit_bike_share_analysis.py @snowflake_native_apps_package.
my_schema.my_stage/streamlit overwrite=true auto_compress=false;
PUT file:///<path_to_your_root_folder>/native_apps/readme.md @
snowflake_native_apps_package.my_schema.my_stage overwrite=true
auto_compress=false;
```

3. Run the following command in the worksheet to verify whether the file upload was successful:

```
List @snowflake_native_apps_package.my_schema.my_stage;
```

After running the command, the uploaded files are displayed in the output:

	name	size	md5	...	last_modified
1	my_stage/manifest.yml	96	542d56aed7857ecfb30564af1bdd9fce		Fri, 5 Apr 2024 13:47:57 GMT
2	my_stage/readme.md	176	f4ca907709c5174a7ffa8c2ffb8e8f76		Fri, 5 Apr 2024 13:47:57 GMT
3	my_stage/scripts/setup.sql	432	b0e7dbc4a1d8b66075da12a058510e19		Fri, 5 Apr 2024 14:52:12 GMT
4	my_stage/streamlit/streamlit_bike_share_analysis.py	17408	b01fd18a9493b12422518e249c6f34d3		Fri, 5 Apr 2024 14:40:04 GMT

Figure 7.14 – Checking the application files via SnowSQL

We can also check the application files in Snowsight:

SNOWFLAKE_NATIVE_APPS_PACKAGE / MY_SCHEMA / MY_STAGE   ···   + Files

🗂 Internal Stage   👤 ACCOUNTADMIN   🕐 2 hours ago

Stage Files   Stage Details

MY_STAGE (4 Files)   🔍 Search   • COMPUTE_WH   ⟳

NAME	SIZE	LAST MODIFIED ↓	
📁 scripts	432.0B	46 minutes ago	···
📁 streamlit	17.0KB	59 minutes ago	···
📄 manifest.yml	96.0B	1 hour ago	···
📄 readme.md	176.0B	1 hour ago	···

Figure 7.15 – Checking the application files via Snowsight

Now that we have uploaded the files to the stage, we will install the application package to Snowflake.

Installing the application

Execute the following command to install the application:

```
CREATE APPLICATION bike_share_native_app
  FROM APPLICATION PACKAGE snowflake_native_apps_package
  USING '@snowflake_native_apps_package.my_schema.my_stage';
```

This automatically creates a new and current version of the application inside Snowflake. The native apps also provide the ability to version the application and update the version, which we will cover in the following section.

↳ Results   ∿ Chart

	status	
1	Application 'BIKE_SHARE_NATIVE_APP' created successfully.	···

Figure 7.16 – Application installed

Now that the application is installed, let's update the version of the application.

Versioning the application

In this section, we will update the application's version and install it on Snowflake. We will utilize the ALTER APPLICATION PACKAGE command to update the previously created application. To update a new version, follow these steps:

1. Run the following command to append a version to SNOWFLAKE_NATIVE_APPS_PACKAGE:

    ```
    ALTER APPLICATION PACKAGE snowflake_native_apps_package
      ADD VERSION v1_0
      USING '@ snowflake_native_apps_package.my_schema.my_stage';
    ```

    This will display the following output:

    Figure 7.17 – Modifying the application version

    This command modifies the application package by incorporating a version derived from the application files uploaded to the named stage in an earlier section.

2. Next, confirm the successful update of the version by executing the following command:

    ```
    SHOW VERSIONS IN APPLICATION PACKAGE snowflake_native_apps_
    package;
    ```

    This will display the updated version as the output:

    Figure 7.18 – The updated version of the application

    This command provides additional insights into the version, including creation timestamp and review status.

3. Before you install the new version, drop the existing application by running the following command:

    ```
    DROP APPLICATION bike_share_native_app;
    ```

You should see the following confirmation message:

Figure 7.19 – Application deleted

Now that the existing application has been dropped, we can install the new application.

4.   Finally, to install the new version of the application, run the following code:

```
CREATE APPLICATION bike_share_native_app
    FROM APPLICATION PACKAGE snowflake_native_apps_package
    USING VERSION V1_0;
```

This creates a new application based on the specified version, ensuring your application incorporates all the latest enhancements and features.

> **Note**
>
> The value provided for VERSION serves as a label rather than a numerical or string value. It's important to note that the patch number is automatically set to 0 when adding a new version. Subsequent patches added to the version will be automatically incremented. However, when introducing a new version, such as V1_1, the patch number for that specific version is reset to 0.

We have successfully deployed the new application version. In the next section, we will run the application to test it within the Snowflake environment.

Testing the application

In this section, we will run our native application in Snowpark to perform the testing. While previous sections primarily utilized SQL statements for testing and information retrieval, Snowsight offers an alternative interface for interacting with and inspecting your application. Moreover, you can examine the Streamlit application you have deployed. To access and explore your application in Snowsight, follow these steps:

1.   In Snowsight, switch to the **ACCOUNTADMIN** role that grants you the necessary privileges to effectively view and manage applications within Snowsight:

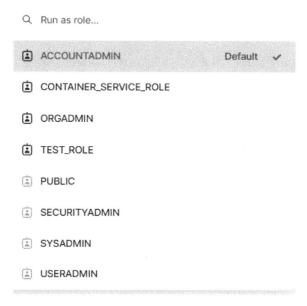

Figure 7.20 – Switching to the ACCOUNTADMIN role

2. Within Snowsight's intuitive navigation menu, locate and select the **Data Products** option, followed by **Apps**. This action directs you to a repository of installed applications within your Snowflake environment:

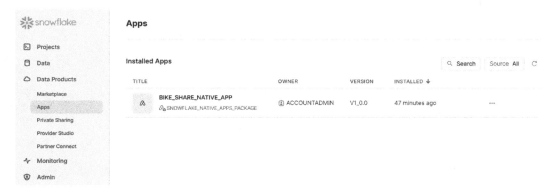

Figure 7.21 – The native applications repository

3. Locate your application within the **Installed Apps** list and select BIKE_SHARE_NATIVE_APP. This action offers a comprehensive view of your application, allowing you to delve into its various components and functionalities. The **Readme** tab provides access to the content you previously added to the README file, offering insights into the purpose and features of your application:

Figure 7.22 – Native application README file

4. Locate and select BIKE_SHARE_STREAMLIT to access your deployed Streamlit application. Upon selection, the content of the SNOWFLAKE_NATIVE_APPS_PACKAGE database is showcased within a Streamlit DataFrame, facilitating dynamic interaction and visualization of your data:

Figure 7.23 – The native application

> **Note on step 4**
>
> Before executing step 4, make sure all the commands under the *CREATE VIEW ON BSD TRAIN TABLE* section in chapter_7.sql are executed. This is necessary to access the BSD_TRAIN table residing in the SNOWPARK_DEFINITIVE_GUIDE database.

The Streamlit application runs successfully within the Snowflake environment, highlighting data representation and interaction features.

In the next section, we will look at how to publish the native application.

Publishing the native application

In this section, we will examine publishing the application by crafting a private listing, utilizing the application package as its core data content. This process enables seamless sharing of your application with other Snowflake users, empowering them to install and use it within their accounts. We will start by setting the release directive.

Setting the default release directive

Before proceeding with the listing creation, it's essential to establish a default release directive specifying which version of your application is accessible to consumers. To designate version `v1_0` and patch `0` as the default release directive, execute the following command:

```
SHOW VERSIONS IN APPLICATION PACKAGE snowflake_native_apps_package;
```

This will display the following output:

	version	patch	label	comment	...	created_on	dropped_on	log_level	trace_level
1	V1_0	0	null	null		2024-04-05 08:10:18.274 -0700	null	OFF	OFF

Figure 7.24 – Checking the application version

The output confirms the successful establishment of the default release directive, ensuring clarity and consistency in version management.

In the next section, we will create the listing for the application.

Creating a listing for your application

With the default release directive in place, we will create a listing for your application, incorporating the application package as its shared data content. Follow these steps to create a listing:

1. Open **Provider Studio** by navigating to **Data Products | Provider Studio** to access the listing creation interface:

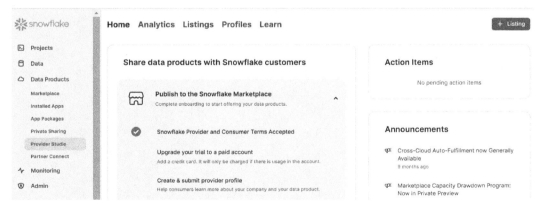

Figure 7.25 – Provider Studio

2. Next, initiate the listing creation by selecting **+ Listing** to open the **Create Listing** window, prompting you to specify the listing's name and other essential details:

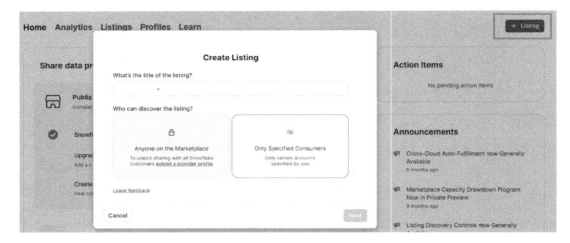

Figure 7.26 – The Create Listing dialog box

3. Enter a descriptive name for your listing and choose the **Only Specified Consumers** visibility option to ensure private sharing with specific accounts only. Select the application package associated with the listing, effectively linking it as the core data content. Provide a comprehensive description of your listing to inform potential users about its functionality and purpose:

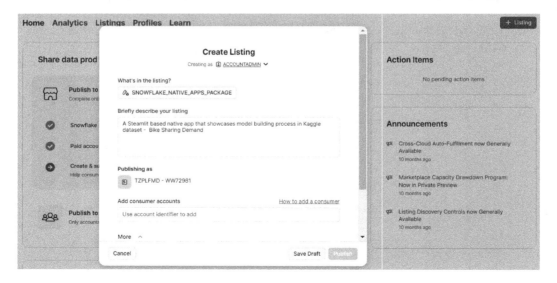

Figure 7.27 – Creating a listing description

4. Add the consumer account by including the account identifier for the account you're utilizing to test the consumer experience of installing the application from the listing:

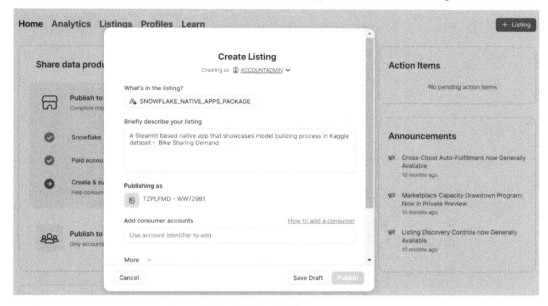

Figure 7.28 – Adding the Consumer account

Now you have successfully curated a private listing containing your application package as the shared data content, facilitating streamlined dissemination and installation of your application among designated users.

The following section will discuss how to manage the native application.

Managing the native application

Configuring and managing installed applications created with the Native App Framework involves various tasks, including viewing installed applications, accessing README information, granting application roles to account roles, and uninstalling applications. Let's delve into each aspect in more detail.

Viewing installed applications

To access and view installed applications or Streamlit apps associated with your account, follow the following steps:

1. From the navigation menu, select **Data Products** and **Apps** to access the list of installed applications:

Figure 7.29 – The list of installed applications

2. Review the list of installed applications, then select the desired application. If you choose an application, the app interface will be displayed:

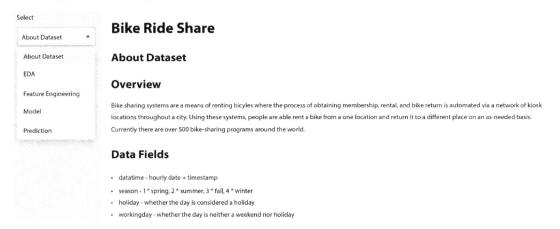

Figure 7.30 – The Streamlit application

Next, we will cover how to view README for the application.

Viewing README for applications

The README provides a meaningful description and other details related to the application. Select the application from the list to access the README, then click on the **Readme** icon in the toolbar to view the README associated with the application:

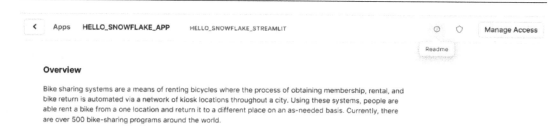

Figure 7.31 – The Streamlit README information

Next, we will cover managing access to the application by granting it to roles.

Managing access to the application

To grant an application role access, select the application and select **Manage Access**. Click on **Add roles**, then select the account role to which you want to grant access to an application role. Click **Done** to confirm:

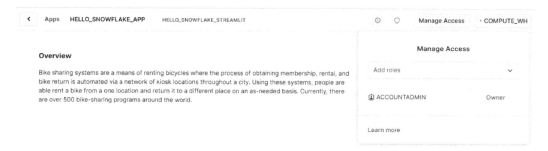

Figure 7.32 – Application access

Alternatively, you can grant application roles to account roles in the consumer account by running the following command:

```
GRANT APPLICATION ROLE bike_share_native_app.app_public
    TO ROLE <ACCOUNT ROLE>;
```

In the final section, we will learn how to uninstall an installed application.

Removing an installed application

The application can be removed using the Snowsight UI or the command line. To remove the application using Snowsight, click on the ellipsis (...) next to the application and choose **Uninstall**. Confirm the action by selecting **Uninstall**:

Figure 7.33 – Uninstalling an application

We can also uninstall it via command by executing the following in the worksheet:

```
DROP APPLICATION BIKE_SHARE_NATIVE_APP;
```

This concludes the management of native applications. Snowflake provides an easy way to do this. We will conclude this chapter with a summary.

Summary

In this chapter, we explored Native Apps within the Snowflake ecosystem, uncovering their dynamic landscape and intrinsic value. The chapter commenced with an insightful introduction to Native Apps in Snowflake, where we delved into their foundational aspects, including robust security features and their significance in the broader native apps landscape. We explored Streamlit in Snowflake, a powerful tool for developing immersive data science and machine learning applications. We provided a comprehensive overview of Streamlit, highlighting its intuitive nature and seamless integration with Snowflake. We navigated through the process of deploying the first Streamlit application within Snowflake, empowering readers to harness this powerful combination for data-driven insights.

Overall, this chapter provided a comprehensive overview of Native Apps in Snowflake, empowering you with the knowledge and tools needed to harness the full potential of this innovative technology within your data workflows. In the next chapter, we will discuss Snowpark Container Services.

8

Introduction to Snowpark Container Services

Containers represent a contemporary method for packaging code in diverse languages, ensuring seamless portability and consistency across various environments. This is particularly true for advanced AI/ML models and comprehensive data-centric applications. These modern data products often handle vast volumes of proprietary data, presenting challenges in efficiently creating, developing, and scaling workloads.

Developers and data scientists often spend more time managing computing resources and clusters than addressing core business challenges. With its unique features, Snowpark Container Services offers a seamless solution to this problem. It allows applications and **large language models (LLMs)** to be executed on containers directly within the Snowflake Data Cloud, reducing the time and effort spent on resource management. This chapter will help you learn about deploying apps and LLMs on containers within Snowpark.

In this chapter, we are going to cover the following topics:

- Introduction to Snowpark Container Services
- Setting up Snowpark Container Services
- Setting up a Snowpark Container Service job
- Deploying LLMs with Snowpark

Technical requirements

To set up the environment, please refer to the technical requirements in the previous chapter. Docker Client and Desktop are also required; you can install Docker from `https://docs.docker.com/get-docker/`.

We'll also be using the Hugging Face API. To obtain the Hugging Face API token, sign up at `https://huggingface.co/`.

The supporting materials are available at `https://github.com/PacktPublishing/The-Ultimate-Guide-To-Snowpark`.

Introduction to Snowpark Container Services

Snowpark Container Services represents a comprehensive managed container solution tailored to facilitate the deployment, management, and scaling of containerized applications within the Snowflake environment. Users can experience the convenience of executing containerized workloads directly within Snowflake, eliminating the need to transfer data outside the Snowflake ecosystem for processing. Snowpark Container Services introduces an **Open Container Initiative** (**OCI**) runtime execution environment meticulously optimized for Snowflake, which empowers users to flawlessly execute OCI images while leveraging the robust capabilities of Snowflake's data platform.

Snowpark Container Services extends Snowpark's capability, empowering developers with a trusted and familiar environment to process non-SQL code seamlessly within Snowflake's governed data domain. This enables applications to effortlessly perform tasks such as connecting to Snowflake, executing SQL queries within a Snowflake virtual warehouse, accessing data files in a Snowflake stage, and processing data with Snowpark models. This streamlined integration fosters an environment conducive to efficient collaboration and focused development efforts within teams.

Developers can create containers tailored to their needs that offer configurable hardware options, including GPU support, enabling a wide range of AI/ML and application workloads within Snowflake through Snowpark. For instance, data science teams can expedite ML tasks by leveraging Python libraries for training and inference while executing resource-intensive generative AI models such as LLMs. App developers can craft and deploy user interfaces using popular frameworks, and data engineers can execute optimized logic within the same processing engine handling SQL or Python DataFrame operations.

In the next section, we will understand how data security works in Snowpark Container Services.

Note on Snowpark Container Services

At the time of writing this chapter, Snowpark Container Services are currently in a private preview phase. Please note that once they become available to all users, there may be slight variations in the API methods compared to what is described in this book. We encourage you to monitor the book's GitHub repository for any new changes and updates to the code contents: `https://github.com/PacktPublishing/The-Ultimate-Guide-To-Snowpark`

Data security in Snowpark Container Services

Snowpark Container Services facilitates the secure deployment of full-stack applications, LLMs, and other advanced data products directly within the data environment. This new runtime option under Snowpark streamlines the deployment, management, and scaling of containerized workloads, including jobs, services, and service functions, leveraging Snowflake-managed infrastructure with customizable hardware configurations, such as GPUs. By adopting this innovative runtime, users can bypass the complexities of managing compute resources and container clusters, allowing seamless integration of sophisticated AI/ML models and applications without compromising data security. With containers operating within the Snowflake environment, there's no need to transfer governed data outside of Snowflake, minimizing exposure to potential security risks. This ensures a secure and robust ecosystem for leveraging internally developed solutions or third-party offerings, such as Snowflake Native Apps, accessible through the Snowflake Marketplace.

In the next section, we will look at the components of Snowpark Containers.

Components of Snowpark Containers

Snowpark Container Services offers a streamlined and fully managed approach to the life cycle management of containerized applications and AI/ML models. Unlike other solutions, it provides a cohesive solution that necessitates piecing together disparate components such as container registries, management services, and computing platforms. Consolidating these elements eliminates the burden of managing computing resources and clusters, thereby accelerating the development and deployment of data applications.

Moreover, Snowpark Container Services simplifies container hosting and deployment by offering a combination of simplicity and scalability. Developers only need to provide their containers, and Snowflake handles the hosting and scaling without requiring extensive knowledge of Kubernetes. Developers can interact with the service using SQL, CLI, or Python interfaces, catering to diverse preferences and workloads. Snowpark Containers has two distinct execution options to accommodate various application requirements: services jobs through using service function, and compute pools. The following diagram shows the different components:

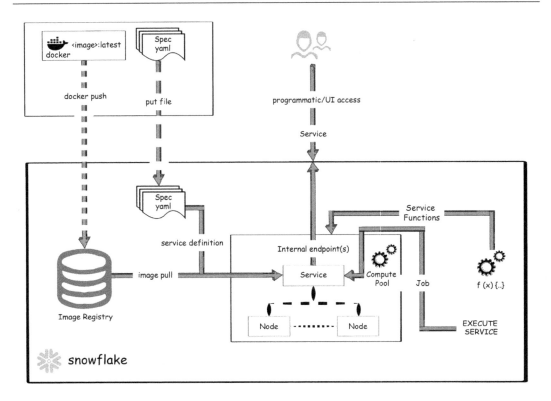

Figure 8.1 – Snowpark Container components

Let's look at each of the options:

- **Services**: A service in Snowflake operates continuously, much like a web service, until explicitly terminated. These services are hosted on secure ingress endpoints and typically host application frontends or APIs. They remain continuously available to handle on-demand requests.

- **Jobs**: These are processes with specific time limits, often initiated manually or scheduled regularly. They encompass various tasks, such as launching container images for machine learning training on GPUs or executing steps within a data pipeline using diverse languages, frameworks, or libraries encapsulated in containers.

- **Service functions**: Functions are time-limited processes designed to receive input, execute specific actions, and be triggered repeatedly by events, leveraging your containerized environments.

- **Compute pools**: A compute pool comprising one or more **virtual machine** (**VM**) nodes serves as the infrastructure upon which Snowflake executes your jobs and services.

Snowpark Container Services also enables developers to deploy applications directly within their end customers' Snowflake accounts using the aforementioned components. This allows them to securely install and operate state-of-the-art offerings, such as hosted notebooks and LLMs, within their Snowflake environment, safeguarding the provider's intellectual property.

In the next section, we will cover how to set up Snowpark Container Services.

Setting up Snowpark Container Services

In this section, we'll lay down the groundwork necessary for exploring Snowpark Container Services. We will use Docker to create an OCI-compliant image to deploy to Snowpark. We'll start by creating Snowflake objects.

Creating Snowflake objects

To create Snowflake objects, follow these steps in Snowsight with the **ACCOUNTADMIN** role:

1. Create a role named `test_role` using the following command. This role will be used for our Snowpark application:

    ```
    USE ROLE ACCOUNTADMIN;
    CREATE ROLE test_role;
    ```

    This will print the following output:

Figure 8.2 – A Snowflake role

2. Create a database and grant access to the database role by running the following command:

    ```
    CREATE DATABASE IF NOT EXISTS SNOWPARK_DEFINITIVE_GUIDE;
    GRANT OWNERSHIP ON DATABASE SNOWPARK_DEFINITIVE_GUIDE
        TO ROLE test_role COPY CURRENT GRANTS;
    ```

This will display the following output:

Figure 8.3 – Granting access

3. We will be granting access to a warehouse for this role by executing the following command:

    ```
    GRANT USAGE ON WAREHOUSE COMPUTE_WH TO ROLE test_role;
    ```

4. Next, we will create a security integration for Snowflake services to access the resources securely by running the following command:

    ```
    CREATE SECURITY INTEGRATION IF NOT EXISTS snowservices_ingress_
    oauth
      TYPE=oauth
      OAUTH_CLIENT=snowservices_ingress
      ENABLED=true;
    ```

    The output is as follows:

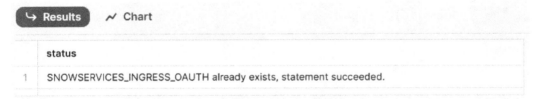

Figure 8.4 – Security integration

5. Next, we will bind the service endpoint on the account to this role by running the following command. This allows access to the service endpoint from the public ingress:

    ```
    GRANT BIND SERVICE ENDPOINT ON ACCOUNT TO ROLE test_role;
    ```

This will display the following output:

6. Finally, we will create a compute pool and assign it to the role by running the following command:

```
CREATE COMPUTE POOL snowpark_cs_compute_pool
MIN_NODES = 1
MAX_NODES = 1
INSTANCE_FAMILY = CPU_X64_XS;
GRANT USAGE, MONITOR ON
  COMPUTE POOL snowpark_cs_compute_pool TO ROLE test_role;
```

This will display the following output:

Figure 8.6 – A compute pool

7. We have now created a role, test_role, and the necessary Snowflake objects we will use for the container services. Now, grant the role to the user you are logged into by running the following command:

```
GRANT ROLE test_role TO USER <user_name>;
```

8. Now that we have the role configured and ready to use, let's create the necessary database-scoped objects:

- Select the database by running the following command:

```
USE DATABASE SNOWPARK_DEFINITIVE_GUIDE;
```

- Create a schema named MY_SCHEMA by running the following code:

```
CREATE SCHEMA IF NOT EXISTS MY_SCHEMA;
```

9. Create an image repository that stores the container image by running the following command:

```
CREATE IMAGE REPOSITORY IF NOT EXISTS snowpark_cs_repository;
```

Once the image is created, you'll see the following output:

	status
1	Image Repository SNOWPARK_CS_REPOSITORY successfully created.

Figure 8.7 – An image repository

10. Finally, create a stage that is used to upload the files by running the following command:

```
CREATE STAGE IF NOT EXISTS snowpark_cs_stage
    DIRECTORY = ( ENABLE = true );
```

You'll see the following output:

	status	⋮
1	Stage area SNOWPARK_CS_STAGE successfully created.	

Figure 8.8 – Stage creation

We will be using the HuffPost dataset, which is available on Kaggle. The dataset is provided in our code repository. The dataset delineates approximately 200,000 headlines from 2012 through May 2018, with an additional 10,000 from May 2018 to 2022, reflecting adjustments in the website's operational dynamics.

In the next section, we will set up the services.

Setting up the services

Flask is a lightweight web framework that allows developers to easily create web applications in Python. It is designed to be flexible, modular, and easy to use, making it a popular choice for building web applications of all sizes. Flask is particularly well-suited for building small to medium-sized web applications, as it provides just enough functionality to get the job done without adding unnecessary complexity.

Flask is used for a wide range of applications, including building web APIs, developing microservices, and creating simple web applications. Its flexibility and simplicity make it a popular choice for developers who want to quickly prototype and deploy web applications. Additionally, Flask can be easily extended with a variety of third-party libraries and tools, making it a powerful and versatile tool for building web applications in Python.

We will be utilizing Flask to write our service code that runs a persisting service to take HTTPS calls.

> **Note on filter service**
>
> Filter service, which we are discussing in the next section, is just one simple example since our focus is more on explaining how to set up Snowpark Container Services rather than building a complex application. By following similar steps, any other use case can be developed.

Setting up the filter service

In this section, we will set up a service called `filter_service`, which filters the table based on a unique ID. We will perform the following steps to set up the service.

Service code

You'll find a Python application encompassing the code for crafting the filter service on the code repository. To initiate, download the provided zip file to a designated directory. Upon download completion, proceed to extract its contents. You'll encounter a `service` directory containing the service code within the extracted files. The directory consists of the Docker file, `filter_service.py`, and the templates for the UI.

Filter Service in Python

The following Python script includes the core logic of our service, encapsulating a minimalistic HTTP server based on Flask, and designed to filter the table based on input. It serves a dual purpose: handling filter requests from Snowflake service functions and furnishing a web UI for submitting filter requests.

```
@app.post("/filter")
def udf_calling_function():
    message = request.json
    logger.debug(f'Received request: {message}')
    if message is None or not message['data']:
        logger.info('Received empty message')
        return {}
    unique_id = message['data']
```

The filter function facilitates communication between a Snowflake service function and the service. This function is adorned with the `@app.post()` decoration, signifying its capability to handle HTTP POST requests directed to the `/filter` path. Upon receiving such requests, the function processes and sends back the filter results encapsulated within the request body:

```
def ui():
    '''
    Main handler for providing a web UI.
    '''
```

```
if request.method == "POST":
    # Getting input in HTML form
    input_text = request.form.get("input")
```

The UI function segment orchestrates a web form presentation and manages filter requests submitted via the web form. Decorated with the `@app.route()` decorator, this function is designated to handle requests targeting the `/ui` path. Upon receiving an HTTP GET request for this path, the server delivers a simple HTML form prompting the user to input a string. Subsequently, upon form submission, an HTTP POST request is dispatched, and the server processes it, returning the original string encapsulated within an HTTP response:

```
@app.get("/healthcheck")
def readiness_probe():
    return "I'm ready!"
```

The `readiness_probe` function, adorned with the `@app.get()` decorator, is primed to handle requests directed to `/healthcheck`. This function is pivotal for Snowflake to verify the service's readiness. When Snowflake initiates a container, it dispatches an HTTP GET request to this path as a health probe, ensuring that only healthy containers handle incoming traffic. The function's implementation is flexible, accommodating various actions to ascertain the service's readiness.

Next, we will look at the Dockerfile in the directory.

The Dockerfile

The Dockerfile serves as a blueprint for constructing an image using Docker. It includes directives on installing the Flask library within the Docker container. The Dockerfile consists of the following:

```
ARG BASE_IMAGE=python:3.10-slim-buster
FROM $BASE_IMAGE
COPY filter_service.py ./
COPY templates/ ./templates/
RUN pip install --upgrade pip && \
    pip install flask && \
    pip install snowflake-snowpark-python[pandas]
CMD ["python3", "filter_service.py"]
```

The code within `filter_service.py` relies on Flask to efficiently handle HTTP requests.

Next, we will examine the UI templates.

UI templates

The UI template files are located at `/template/basic_ui.html`. They render a web form for the filter service's publicly exposed endpoint. This form is displayed when the public endpoint URL is loaded in a web browser with `/ui` appended. Users can input a string via this form, and upon submission, the service filters the table with submitted row given as string within an HTTP response.

In the next section, we will cover the service function.

The service function

A service function serves as a conduit for communicating with your service. A **user-defined function (UDF)** is tethered to a service endpoint. Upon execution, the service function dispatches a request to the associated service endpoint and awaits a response. Creating such service functions involves executing the CREATE FUNCTION command with specified parameters, such as the `filter_doc_udf` function:

```
CREATE FUNCTION filter_doc_udf (InputText varchar)
   RETURNS varchar
   SERVICE=filter_service
   ENDPOINT=filterendpoint
   AS '/filter';
```

This function, for instance, accepts a string as input and returns a string, with the SERVICE property designating the service (`filter_service`) and the ENDPOINT property specifying the user-friendly endpoint name (`filterendpoint`). The AS `'/filter'` designation denotes the path for the service, tying it to the corresponding function within `filter_service.py`. Thus, invoking this function triggers Snowflake to dispatch a request to the designated path within the service container.

In the next section, we will build the Docker image.

Building the Docker image

In this section, we will construct the image using the Linux/AMD64 base, which is compatible with Snowpark, and dispatch it to your account's image repository. To build the Docker image, perform the following steps:

1. Obtain the repository URL by executing the following SQL command:

    ```
    SHOW IMAGE REPOSITORIES;
    ```

    This will display all the image repositories:

Figure 8.9 – The image repositories

The **repository_url** column in the output furnishes the essential URL, and the hostname delineated in the repository URL denotes the registry hostname.

2. The following commands require Docker Desktop to be installed in the system. You can install it from https://www.docker.com/products/docker-desktop/ before proceeding with the commands. Next, in the local terminal window, switch to the service directory containing the unzipped files and execute the subsequent docker build command using the Docker CLI:

```
docker build --rm --platform linux/amd64 -t <orgname>-
<acctname>.registry.snowflakecomputing.com/snowpark_definitive_
guide/my_schema/snowpark_cs_repository/my_filter_service_
image:latest .
```

This command designates the current working directory (.) as the path for building the latest image from the Docker file. The output will be as follows:

Figure 8.10 – The Docker build command

3. Next, we'll authenticate Docker with Snowflake. To authenticate Docker with the Snowflake registry, execute the following command:

```
docker login <registry_hostname> -u <username>
```

Specify your Snowflake username for the username parameter. Docker will prompt you for your password. Use the Snowflake password to authenticate:

```
Password:
Login Succeeded
```

Figure 8.11 – Repository login

4. Finally, upload the image to the Docker registry by executing the following command:

```
docker push <orgname>-<acctname>.registry.snowflakecomputing.
com/snowpark_definitive_guide/my_schema/snowpark_cs_repository/
my_filter_service_image:latest
```

You should see the following output:

```
The push refers to repository [tzplfmd-ww72981.registry.snowflakecomputing.com/snowpark_definitive_guide/my_schema/snowpark_cs_repositor
y/my_filter_service_image]
abd2067e9cae: Pushed
fe0e7d3f9844: Pushed
ea8b7ce82acf: Pushed
c5321f7f53ff: Pushed
df6c1b185b95: Pushed
b23fedba7dbd: Pushed
ae2d55769c5e: Pushed
e2ef8a51359d: Pushed
latest: digest: sha256:b3ec43c2baa9adc0459ff60fb86f7cb37cc56b78b039f70eecd64e67ba69c5c9 size: 1998
```

Figure 8.12 – Repository push

The image is now available in the registry for deployment into Container Services.

In the next section, we will examine how to deploy the service, but it is always best practice to test your build locally before pushing it to the Snowflake repository. This part is not explained in this section as it is beyond the scope of this book.

Deploying the service

In this section, we'll guide you through deploying the service and establishing a service function to facilitate communication with it. We will start by deploying the service, which requires the existing compute pool. Let's start by checking the compute pool by running the following command:

```
DESCRIBE COMPUTE POOL snowpark_cs_compute_pool;
```

Figure 8.13 – The compute pool

If it's in the **STARTING** state, you'll need to wait until it transitions to **ACTIVE** or **IDLE**.

Now that the pool is active, we can create the service in the next section.

Creating the service

We can create the service by running it using test_role. To do that, run the following command:

```
USE ROLE test_role;
CREATE SERVICE filter_service
  IN COMPUTE POOL snowpark_cs_compute_pool
  FROM SPECIFICATION $$
    spec:
```

```
containers:
- name: filter
  image: /snowpark_definitive_guide/my_schema/snowpark_cs_
repository/my_filter_service_image:latest
  env:
    SERVER_PORT: 8000
  readinessProbe:
    port: 8000
    path: /healthcheck
endpoints:
- name: filterendpoint
  port: 8000
  public: true
$$
MIN_INSTANCES=1
MAX_INSTANCES=1;
```

We are using the image that we have built to deploy the service. The service should be created within Snowflake.

Once the service is created, you can execute the following SQL command to check its status:

```
SELECT SYSTEM$GET_SERVICE_STATUS('filter_service');
```

The output should show that the service is running. The information about the service can be obtained by running the following command:

```
DESCRIBE SERVICE filter_service;
```

This will display the details, as shown in the following screenshot:

Figure 8.14 – Service information

In the next section, we will create the service function.

Creating a service function

The service function performs the filter function and associates it with an endpoint. To create a service function, execute the following command:

```
CREATE FUNCTION filter_doc_udf (InputText varchar)
RETURNS varchar
SERVICE=filter_service
ENDPOINT=filterendpoint
AS '/filter';
```

Here, the SERVICE property links the UDF with the filter_service service, while the ENDPOINT property associates it with the filterendpoint endpoint within the service. The AS '/filter' specification denotes the HTTP path leading to the filter server, which can be located within the service code.

Once the previous SQL statement is executed correctly, you can see the service function you created in Snowsight under **Functions**.

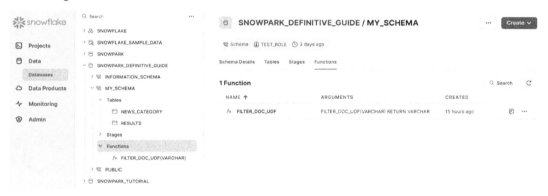

Figure 8.15 – The service function

Now the function is ready to be executed.

Executing the function

We will switch to the context we created earlier in the chapter by running the following command:

```
USE ROLE test_role;
USE DATABASE SNOWPARK_DEFINITIVE_GUIDE;
USE SCHEMA MY_SCHEMA;
USE WAREHOUSE compute_wh;
```

You should get the following confirmation:

Figure 8.16 – Function execution

With the context set up, you can initiate communication with the filter service by invoking the service function within a query. To call the `filter_doc_udf` service function, execute the following SELECT statement, providing a sample input string (`'122880'`):

```
SELECT filter_doc_udf('122880');
```

Upon executing this query, Snowflake dispatches a POST request to the service endpoint (filterendpoint). Upon receiving the request, the service utilizes the input string to filter the table for UNIQUE_ID and sends back the appropriate row in the response, as shown here:

Figure 8.17 – The filter function

The service exposes its endpoint publicly but still securely behind the Snowflake authentication mechanism, as specified in the inline specification provided within the CREATE SERVICE command. Consequently, you can access a web UI that the service exposes to the internet and send requests to the service from a web browser. To find the URL of the public endpoint the service exposes, execute the following command:

```
SHOW ENDPOINTS IN SERVICE filter_service;
```

To access the web UI, append /ui to the endpoint URL and paste it into the web browser. This action triggers the execution of the ui() function specified in the filter_service.py script:

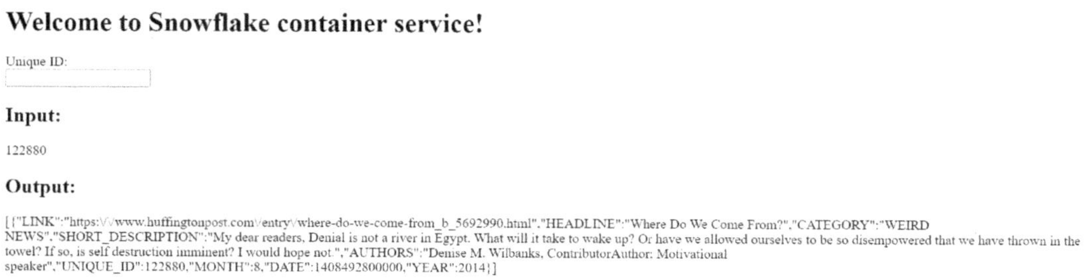

Welcome to Snowflake container service!

Unique ID:

Input:

122880

Output:

[{"LINK":"https://www.huffingtonpost.com/entry/where-do-we-come-from_b_5692990.html","HEADLINE":"Where Do We Come From?","CATEGORY":"WEIRD NEWS","SHORT_DESCRIPTION":"My dear readers. Denial is not a river in Egypt. What will it take to wake up? Or have we allowed ourselves to be so disempowered that we have thrown in the towel? If so, is self destruction imminent? I would hope not.","AUTHORS":"Denise M. Wilbanks, ContributorAuthor; Motivational speaker","UNIQUE_ID":122880,"MONTH":8,"DATE":1408492800000,"YEAR":2014}]

Figure 8.18 – The service UI

Please note that the first time you access the endpoint URL, you'll be prompted to log in to Snowflake. Ensure you log in as the same user who created the service to guarantee you possess the necessary privileges.

We have successfully deployed the service and the components. In the next section, we will look at the service job.

Setting up a Snowpark Container Service job

In this section, we will create a simple job to connect to a Snowflake table and conduct some feature engineering tasks by generating new columns. Subsequently, we'll save the resultant data to the same table within the Snowflake environment. Unlike services, jobs are short-lived, providing a one-time execution of tasks.

In the next section, we will set up the container job.

Setting up the job

For the job, instead of the Flask server implementation for services, we'll utilize a straightforward `main.py` file to execute the job action. We will perform the following steps to set up the job.

Job code

The code for this section is in our GitHub repository under the `chapter_8` folder. The folder contains the following files, which are required for the job.

The main.py file

The `main.py` file is the core Python script for orchestrating the job's execution. At its heart lies the following `run_job()` function, invoked when the script is executed. This function plays a pivotal role in reading environment variables and utilizing them to set default values for various parameters that are essential for connecting to Snowflake.

```
def run_job():
    """
```

```
Main body of this job.
"""
logger = get_logger()
logger.info("Job started")

# Parse input arguments
args = get_arg_parser().parse_args()
table = args.table
column = args.date_column
```

While Snowflake automatically populates some parameters when the image runs within its environment, explicit provision is required when testing the image locally. The run_job() function gets a table name and column to perform feature engineering from the spec.

The Dockerfile

The Dockerfile encapsulates all the necessary commands required to build an image using Docker. This file resembles what we've previously implemented in our service section, ensuring consistency and coherence across different Snowpark Container Services environment components.

The job specification file

The following job specification file provides Snowflake with essential container configuration information. Snowflake leverages the information provided in the my_job_spec.yaml specification file to configure and execute your job seamlessly. In addition to mandatory fields such as container. name and container.image, this specification file includes optional fields such as container. args, which list the arguments required for job execution.

```
spec:
  container:
  - name: main
    image: /snowpark_definitive_guide/my_schema/snowpark_cs_
repository/my_job_image:latest
    env:
      SNOWFLAKE_WAREHOUSE: compute_wh
    args:
    - "--table=NEWS_CATEGORY"
    - "--date_column=DATE"
```

Notably, the --query argument specifies the query to be executed when the job runs, while the --result_table argument identifies the table where the query results will be stored.

In the next section, we will deploy the job.

Deploying the job

To upload your job specification file (my_job_spec.yaml) into the Snowflake environment, you have a couple of options for uploading it to the designated stage:

- **Snowsight web interface**: Utilizing the Snowsight web interface offers a user-friendly approach to uploading your job specification file. Following the instructions we have covered in previous chapters, you can effortlessly navigate the process and ensure successful integration.

- **SnowSQL command-line interface (CLI)**: Alternatively, you can use the SnowSQL CLI to execute the file upload process by executing the following PUT command syntax:

  ```
  PUT file://path/to/my_job_spec.yaml @snowpark_cs_stage/my_job_
  spec.yaml OVERWRITE=TRUE;
  ```

  Upon successfully executing the PUT command, detailed information regarding the uploaded file will be displayed in Snowsight:

Figure 8.19 – Job upload

Now that the job file has been uploaded, we will execute the job in the next section.

Executing the job

To kick off the execution of a job, you'll utilize the EXECUTE SERVICE command, which acts as the catalyst for launching the specified task. Run the following command to trigger the job (this command may change since we are in private preview at the time of writing):

```
EXECUTE SERVICE IN COMPUTE POOL snowpark_cs_compute_pool
  FROM @snowpark_cs_stage SPEC='my_job_spec.yaml';
```

Alternatively, you can use the following:

```
EXECUTE JOB SERVICE
  IN COMPUTE POOL snowpark_cs_compute_pool
  NAME = test_job
  FROM @SNOWPARK_CS_STAGE
  SPECIFICATION_FILE='my_job_spec.yaml';
```

The specified compute pool, `snowpark_cs_compute_pool`, determines the allocation of computational resources necessary for the job's successful execution. The `@snowpark_cs_stage` notation denotes the designated stage within Snowflake where the job specification file is stored, facilitating seamless access to the required configuration details. The `my_job_spec.yaml` file refers to the specific configuration file containing the instructions and parameters for executing the job seamlessly. Successful execution of the command should display the following output:

Figure 8.20 – Job execution

Upon execution, the job performs the specified SQL statement and saves the resultant data to a designated table, as outlined within the job specification file (`my_job_spec.yaml`). It's crucial to note that the execution of the SQL statement does not occur within the Docker container itself. Instead, the container connects with Snowflake, leveraging a Snowflake warehouse to execute the SQL statement efficiently. The `EXECUTE SERVICE` command returns the output containing vital information, including the Snowflake-assigned **UUID** (short for **Universally Unique Identifier**) of the job. This UUID serves as a unique identifier for the executed job, aiding in tracking and monitoring its progress and status.

In the next section, we will deploy an LLM for Snowpark Container Services.

Deploying LLMs with Snowpark

Modern enterprises increasingly demand that LLMs be harnessed with proprietary data. Open source and proprietary models play pivotal roles in enabling this transition. However, the main challenge is finding a robust platform capable of effectively leveraging LLMs' power. Snowflake empowers organizations to apply near-magical generative AI transformations to their data. By leveraging advanced LLM models within Snowflake, organizations can efficiently operate with large volumes of data, enabling generative AI use cases. In this section, we will discuss deploying LLM models within Snowpark Container Services.

In this walk-through, we'll explore how to harness publicly accessible data to demonstrate the transformative capabilities of Snowflake's ecosystem by deploying the Llama 2 LLM from the Hugging Face repository.

> **Note**
>
> Llama 2 by Meta, housed within Hugging Face's library, epitomizes advanced **natural language processing** (**NLP**) technology. As stipulated by Meta's specific terms of service, you'll need a Hugging Face token to access Llama 2 with Hugging Face. Please visit `https://huggingface.co/docs/hub/en/security-tokens` to learn more.

Preparing the LLM

We will start by preparing the LLM by utilizing our convenient wrapper around the Hugging Face Transformers API, and harness the capabilities of Llama 2 7B from Hugging Face. To achieve this, run the following code:

```
HF_AUTH_TOKEN = " ************************ "
registry = model_registry.ModelRegistry(session=session, database_
name="SNOWPARK_DEFINITIVE_GUIDE", schema_name="MY_SCHEMA", create_if_
not_exists=True)
llama_model = huggingface_pipeline.HuggingFacePipelineModel(task="te
xt-generation", model="meta-llama/Llama-2-7b-chat-hf", token=HF_AUTH_
TOKEN, return_full_text=False, max_new_tokens=100)
```

Make sure to replace `HF_AUTH_TOKEN` with your token from Hugging Face. The code creates the model registry and assigns the model from the Hugging Face registry. The model is obtained from the Hugging Face registry and directly imported into Snowpark.

Next, we will register the model within Snowpark ML.

Registering the model

Next, we'll utilize the model registry's `log_model` API within Snowpark ML to register the model. This involves specifying a model name and a version string and providing the model obtained in the previous step:

```
MODEL_NAME = "LLAMA2_MODEL_7b_CHAT"
MODEL_VERSION = "1"
llama_model=registry.log_model(
```

```
    model_name=MODEL_NAME,
    model_version=MODEL_VERSION,
    model=llama_model
)
```

You should see an output similar to the following:

```
<snowflake.ml.registry.model_registry.ModelReference at 0x1bdf116e3d0>
```

Figure 8.21 – Model registration

The model is now registered in the registry. Now that the model is ready, we will deploy it to Container Services.

Deploying the model to Snowpark Container Services

Now, let us deploy the model to our designated compute pool. Once the deployment process is initiated, the model will become accessible as a Snowpark Container Services endpoint. Run the following code to deploy the model to Container Services. To run this step, you may need to alter your compute pool to include a GPU instance, or you can create a new compute pool with a GPU instance.

```
llama_model.deploy(
  deployment_name="llama_predict",
  platform=deploy_platforms.TargetPlatform.SNOWPARK_CONTAINER_
SERVICES,
  options={
          "compute_pool": "snowpark_cs_compute_pool",
          "num_gpus": 1
  },
)
"external_access_integrations": ["ALLOW_ALL_ACCESS_INTEGRATION"]
```

This streamlined deployment process highlights how Snowpark ML simplifies the deployment of LLMs, handling the creation of the corresponding Snowpark Container Services SERVICE definition, packaging the model within a Docker image along with its runtime dependencies, and launching the service within the specified compute pool.

After executing the code, you should see a similar output:

```
CPU times: user 792 ms, sys: 166 ms, total: 958 ms
Wall time: 10min 20s
Out[10] <snowflake.ml.registry.model_registry.ModelReference at 0x7f06a9e78dc0>
```

Figure 8.22 – Deploying the LLM model to Snowpark Container Services

In the next section, we will execute this model in the container.

> **Note on model deployment**
>
> Only the snippets required for explanation are shown in this section. The complete code is available in the `chapter_8.ipynb` notebook in GitHub. You should be mindful of the model deployment step as it takes considerable time and resources.

Running the model

Invoke the model by supplying the subset of the NEWS_CATEGORY table with the `inputs` column containing the prompt:

```
res = llama_model_ref.predict( deployment_name=DEPLOYMENT_NAME,
data=input_df )
```

This yields a Snowpark DataFrame with an output column containing the model's response for each row. The raw response intersperses text with the expected JSON output, exemplified as follows:

```
{
  "category": "Art",
  "keywords": [
    "Gertrude",
    "contemporary art",
    "democratization",
    "demystification"
  ],
  "importance": 9
}
```

Deploying and executing an LLM model is very easy with Snowpark Container Services.

We will conclude the chapter with a summary.

Summary

In this chapter, we explored Snowpark Container Services, a powerful solution designed to simplify the deployment and management of containerized applications within the Snowflake ecosystem. We discussed the distinction between jobs and services within Snowpark Container Services, highlighting their respective functionalities and use cases. We demonstrated how to effectively configure, deploy, and manage jobs and services through practical implementation examples.

Additionally, we delved into containerization through Snowpark ML, showcasing how Snowflake users can seamlessly leverage advanced ML models within their environment. By integrating a language model from Hugging Face, we illustrated how Snowpark ML facilitates the integration of containerized models, enabling sophisticated NLP tasks directly within Snowflake. Overall, this chapter equips you with the knowledge and tools to harness the transformative potential of SCS and Snowpark ML in your data-driven initiatives.

In conclusion, Snowpark Container Services offers a compelling value proposition for businesses seeking efficient and scalable data processing solutions. By enabling secure execution of containerized workloads directly within Snowflake, Snowpark eliminates the need for data movement, ensuring data integrity and reducing latency. Additionally, Snowpark simplifies the development and deployment of data applications, allowing teams to focus on innovation rather than infrastructure management. Automated container management further streamlines operational tasks, enhancing overall productivity and agility.

With this, we conclude the book. Thank you for reading.

Index

packtpub.com

Subscribe to our online digital library for full access to over 7,000 books and videos, as well as industry leading tools to help you plan your personal development and advance your career. For more information, please visit our website.

Why subscribe?

- Spend less time learning and more time coding with practical eBooks and Videos from over 4,000 industry professionals

- Improve your learning with Skill Plans built especially for you

- Get a free eBook or video every month

- Fully searchable for easy access to vital information

- Copy and paste, print, and bookmark content

Did you know that Packt offers eBook versions of every book published, with PDF and ePub files available? You can upgrade to the eBook version at packtpub.com and as a print book customer, you are entitled to a discount on the eBook copy. Get in touch with us at customercare@packtpub.com for more details.

At www.packtpub.com, you can also read a collection of free technical articles, sign up for a range of free newsletters, and receive exclusive discounts and offers on Packt books and eBooks.

Other Books You May Enjoy

If you enjoyed this book, you may be interested in these other books by Packt:

Data Modeling with Snowflake

Serge Gershkovich

ISBN: 978-1-83763-445-3

- Discover the time-saving benefits and applications of data modeling
- Learn about Snowflake's cloud-native architecture and its features
- Understand and apply modeling techniques using Snowflake objects
- Universal modeling concepts and language through Snowflake objects
- Get comfortable reading and transforming semistructured data
- Learn directly with pre-built recipes and examples
- Learn to apply modeling frameworks from Star to Data Vault

Data Engineering with dbt

Roberto Zagni

ISBN: 978-1-80324-628-4

- Create a dbt Cloud account and understand the ELT workflow
- Combine Snowflake and dbt for building modern data engineering pipelines
- Use SQL to transform raw data into usable data, and test its accuracy
- Write dbt macros and use Jinja to apply software engineering principles
- Test data and transformations to ensure reliability and data quality
- Build a lightweight pragmatic data platform using proven patterns
- Write easy-to-maintain idempotent code using dbt materialization

Packt is searching for authors like you

If you're interested in becoming an author for Packt, please visit authors.packtpub.com and apply today. We have worked with thousands of developers and tech professionals, just like you, to help them share their insight with the global tech community. You can make a general application, apply for a specific hot topic that we are recruiting an author for, or submit your own idea.

Share your thoughts

Now you've finished *The Ultimate Guide to Snowpark*, we'd love to hear your thoughts! Scan the QR code below to go straight to the Amazon review page for this book and share your feedback or leave a review on the site that you purchased it from.

https://packt.link/r/1-805-12341-6

Your review is important to us and the tech community and will help us make sure we're delivering excellent quality content.

Download a free PDF copy of this book

Thanks for purchasing this book!

Do you like to read on the go but are unable to carry your print books everywhere?

Is your eBook purchase not compatible with the device of your choice?

Don't worry, now with every Packt book you get a DRM-free PDF version of that book at no cost.

Read anywhere, any place, on any device. Search, copy, and paste code from your favorite technical books directly into your application.

The perks don't stop there, you can get exclusive access to discounts, newsletters, and great free content in your inbox daily

Follow these simple steps to get the benefits:

1. Scan the QR code or visit the link below

https://packt.link/free-ebook/9781805123415

2. Submit your proof of purchase

3. That's it! We'll send your free PDF and other benefits to your email directly

www.ingramcontent.com/pod-product-compliance
Lightning Source LLC
LaVergne TN
LVHW081521050326
832903LV00025B/1575